NATIONAL ACADEMIES Sciences Engineering Medicine

NATIONAL ACADEMIES PRESS
Washington, DC

Developing a Strategy to Evaluate the National Climate Assessment

Committee to Develop a Strategy to Evaluate the National Climate Assessment

Board on Atmospheric Sciences and Climate

Division on Earth and Life Studies

Board on Environmental Change and Society
Committee on National Statistics

Division of Behavioral and Social Sciences and Education

Consensus Study Report

NATIONAL ACADEMIES PRESS 500 Fifth Street, NW Washington, DC 20001

This activity was supported by a contract between the National Academy of Sciences and the National Aeronautics and Space Administration. Any opinions, findings, conclusions, or recommendations expressed in this publication do not necessarily reflect the views of any organization or agency that provided support for the project.

International Standard Book Number-13: 978-0-309-72500-2
International Standard Book Number-10: 0-309-72500-3
Digital Object Identifier: https://doi.org/10.17226/27923
Library of Congress Control Number: 2024949739

This publication is available from the National Academies Press, 500 Fifth Street, NW, Keck 360, Washington, DC 20001; (800) 624-6242 or (202) 334-3313; http://www.nap.edu.

Copyright 2024 by the National Academy of Sciences. National Academies of Sciences, Engineering, and Medicine and National Academies Press and the graphical logos for each are all trademarks of the National Academy of Sciences. All rights reserved.

Printed in the United States of America.

Suggested citation: National Academies of Sciences, Engineering, and Medicine. 2024. *Developing a Strategy to Evaluate the National Climate Assessment*. Washington, DC: The National Academies Press. https://doi.org/10.17226/27923.

The **National Academy of Sciences** was established in 1863 by an Act of Congress, signed by President Lincoln, as a private, nongovernmental institution to advise the nation on issues related to science and technology. Members are elected by their peers for outstanding contributions to research. Dr. Marcia McNutt is president.

The **National Academy of Engineering** was established in 1964 under the charter of the National Academy of Sciences to bring the practices of engineering to advising the nation. Members are elected by their peers for extraordinary contributions to engineering. Dr. John L. Anderson is president.

The **National Academy of Medicine** (formerly the Institute of Medicine) was established in 1970 under the charter of the National Academy of Sciences to advise the nation on medical and health issues. Members are elected by their peers for distinguished contributions to medicine and health. Dr. Victor J. Dzau is president.

The three Academies work together as the **National Academies of Sciences, Engineering, and Medicine** to provide independent, objective analysis and advice to the nation and conduct other activities to solve complex problems and inform public policy decisions. The National Academies also encourage education and research, recognize outstanding contributions to knowledge, and increase public understanding in matters of science, engineering, and medicine.

Learn more about the National Academies of Sciences, Engineering, and Medicine at **www.nationalacademies.org**.

Consensus Study Reports published by the National Academies of Sciences, Engineering, and Medicine document the evidence-based consensus on the study's statement of task by an authoring committee of experts. Reports typically include findings, conclusions, and recommendations based on information gathered by the committee and the committee's deliberations. Each report has been subjected to a rigorous and independent peer-review process and it represents the position of the National Academies on the statement of task.

Proceedings published by the National Academies of Sciences, Engineering, and Medicine chronicle the presentations and discussions at a workshop, symposium, or other event convened by the National Academies. The statements and opinions contained in proceedings are those of the participants and are not endorsed by other participants, the planning committee, or the National Academies.

Rapid Expert Consultations published by the National Academies of Sciences, Engineering, and Medicine are authored by subject-matter experts on narrowly focused topics that can be supported by a body of evidence. The discussions contained in rapid expert consultations are considered those of the authors and do not contain policy recommendations. Rapid expert consultations are reviewed by the institution before release.

For information about other products and activities of the National Academies, please visit www.nationalacademies.org/about/whatwedo.

**COMMITTEE TO DEVELOP A STRATEGY TO EVALUATE
THE NATIONAL CLIMATE ASSESSMENT[1]**

KAI N. LEE (*Chair*), Principal, Owl of Minerva, LLC
ANN M. GALLAGHER, Science Education Coordinator, U.S. National Park Service; Ph.D. Candidate, Strome College of Business, School of Public Service, Old Dominion University
MATTHEW O. GRIBBLE, Associate Professor and Associate Chief for Research in Occupational, Environmental, and Climate Medicine, University of California, San Francisco
SCOTT KALAFATIS, Deputy University Director, Northwest Climate Adaptation Science Center
JESSICA KRONSTADT, Program Director, Planetary Health Alliance
GLYNIS C. LOUGH, Affiliate, Aspen Global Change Institute
MICHELLE MIRO, Senior Information Scientist, RAND Corporation
ARIANE PINSON, Research Social Scientist, U.S. Army Corps of Engineers
UROOJ RAJA, Assistant Professor, Loyola University Chicago
CARLOS RODRIGUEZ FRANCO, Senior Forester, U.S. Forest Service
KATHLEEN SEGERSON, Board of Trustees Distinguished Professor of Economics, University of Connecticut
KRISTIN MARIE FISCHER TIMM, Research Assistant Professor, University of Alaska Fairbanks

Study Staff

STEVEN STICHTER, Study Director (*as of March 25, 2024*), Senior Program Officer, Board on Atmospheric Sciences and Climate (BASC)
BRADFORD CHANEY, Senior Program Officer, Committee on National Statistics
LINDSAY MOLLER, Senior Program Assistant, BASC
HUGH WALPOLE, Study Director, Associate Program Officer, BASC (*until March 22, 2024*)

---

[1] All committee members serve as an individual rather than as a representative of a group or organization. The contributions of the committee members do not necessarily reflect the views of their employers or affiliated organizations.

BOARD ON ATMOSPHERIC SCIENCES AND CLIMATE

MARY GLACKIN (*Chair*), The Weather Company, an IBM Business (Retired)
JOSEPH ÁRVAI, University of Southern California
CYNTHIA S. ATHERTON, Heising-Simons Foundation
ELIZABETH A. BARNES, Colorado State University
BRAD R. COLMAN, The Climate Corporation (Retired)
BART E. CROES, California Air Resources Board (Retired)
MINGHUI DIAO, San Jose State University
NEIL DONAHUE, Carnegie Mellon University
LESLEY-ANN DUPIGNY-GIROUX, University of Vermont
EFI FOUFOULA-GEORGIOU (NAE), University of California, Irvine
KEVIN GURNEY, Northern Arizona University
ANDREA LOPEZ LANG, University of Albany
MARIA CARMEN LEMOS (NAS), University of Michigan
ZHANQING LI, University of Maryland
AMY MCGOVERN, Oklahoma State University
LINDA O. MEARNS, National Center for Atmospheric Research
JONATHAN A. PATZ (NAM), University of Wisconsin-Madison
KEVIN REED, Stony Brook University
J. MARSHALL SHEPHERD (NAS/NAE), University of Georgia
ARADHNA TRIPATI, University of California, Los Angeles
BERNADETTE WOODS PLACKY, Climate Central

National Academies Staff

ELIZABETH EIDE, Acting Director
MAGGIE WALSER, Acting Director
KATELYN CREWS, Program Assistant
APURVA DAVE, Senior Program Officer
MORGAN DISBROW-MONZ, Program Officer
KATRINA HUI, Associate Program Officer
ANNIE MANVILLE, Program Assistant
BRIDGET McGOVERN, Program Officer
APRIL MELVIN, Senior Program Officer
LINDSAY MOLLER, Senior Program Assistant
RACHEL SILVERN, Program Officer
STEVEN STICHTER, Senior Program Officer

COMMITTEE ON NATIONAL STATISTICS

KATHARINE G. ABRAHAM (*Chair*), Department of Economics, University of Maryland, College Park
MICK P. COUPER, Institute for Social Research, University of Michigan
WILLIAM "SANDY" DARITY, Sanford School of Public Policy, Duke University
ROBERT M. GOERGE, NORC at the University of Chicago
ERICA L. GROSHEN, School of Industrial and Labor Relations, Cornell University
DANIEL E. HO, Stanford Law School, Stanford University
HILARY W. HOYNES, Goldman School of Public Policy, University of California, Berkeley
H. V. JAGADISH, Michigan Institute for Data Science, University of Michigan
SHARON LOHR, School of Mathematical and Statistical Sciences, Arizona State University, *Emerita*
NELA RICHARDSON, ADP Research Institute, Roseland, NJ
ELIZABETH A. STUART, Department of Mental Health, Johns Hopkins Bloomberg School of Public Health

National Academies Staff

MELISSA CHIU, Senior Board Director
CONNIE CITRO, Senior Scholar
BRADFORD CHANEY, Senior Program Officer
DANIEL CORK, Senior Program Officer
ALEX HENDERSON, Senior Program Assistant
DAVID JOHNSON, Senior Program Officer
REBECCA KRONE, Administrative Coordinator
CHRIS MACKIE, Senior Program Officer
ANTHONY MANN, Program Associate
KRISZTINA MARTON, Senior Program Officer
JENNIFER PARK, Senior Program Officer
SITARA RAHIAB, Senior Program Assistant
KATRINA STONE, Senior Program Officer

BOARD ON ENVIRONMENTAL CHANGE AND SOCIETY

MICHAEL P. VANDENBERGH (*Chair*), Vanderbilt University
BILAL M. AYYUB, University of Maryland, College Park
EDUARDO S. BRONDIZIO, Indiana University Bloomington
LISA DILLING, University of Colorado Boulder
KENNETH GILLINGHAM, Yale University
MARY H. HAYDEN, University of Colorado
LORI HUNTER, University of Colorado Boulder
STEPHEN H. LINDER, University of Texas Health Science Center at Houston
GLEN M. MACDONALD, University of California, Los Angeles
GARY E. MACHLIS, Clemson University
BENJAMIN PRESTON, RAND Corporation
JACKIE QATALIÑA SCHAEFFER, Alaska Native Tribal Health Consortium
JESSE C. RIBOT, American University
MADELINE I. SCHOMBURG, Energy Futures Initiative Foundation
BENJAMIN K. SOVACOOL, Boston University
ADELLE DAWN THOMAS, University of the Bahamas
CATHY L. WHITLOCK (NAS), Montana State University, *Emerita*

National Academies Staff

PATTI SIMON, Acting Director
JOSHUA LANG, Program Coordinator
JOHN BEN SOILEAU, Program Officer
DANIEL TALMAGE, Program Officer

Reviewers

This Consensus Study Report was reviewed in draft form by individuals chosen for their diverse perspectives and technical expertise. The purpose of this independent review is to provide candid and critical comments that will assist the National Academies of Sciences, Engineering, and Medicine in making each published report as sound as possible and to ensure that it meets the institutional standards for quality, objectivity, evidence, and responsiveness to the study charge. The review comments and draft manuscript remain confidential to protect the integrity of the deliberative process.

We thank the following individuals for their review of this report:

THOMAS DIETZ (NAS), Michigan State University (*retired*), Grand Isle, Vermont
KENNETH FRANK, Michigan State University, East Lansing
JOY FRECHTLING, Westat, Rockville, Maryland
CHRISTINE KIRCHHOFF, The Pennsylvania State University, University Park
GARY YOHE, Wesleyan University, Portland, Connecticut

Although the reviewers listed above provided many constructive comments and suggestions, they were not asked to endorse the conclusions or recommendations of this report, nor did they see the final draft before its release. The review of this report was overseen by **KRISTIE L. EBI,** University of Washington, Seattle, and **CHRISTOPHER B. FIELD (NAS),** Stanford University, Stanford, California. They were responsible for making certain that an independent examination of this report was carried out in accordance with the standards of the National Academies and that all review comments were carefully considered. Responsibility for the final content rests entirely with the authoring committee and the National Academies.

Contents

Key Terms Used in This Report   xv

SUMMARY   1

1 INTRODUCTION   9
Evaluations, 10
Statement of Task, 12
The Committee's Approach, 14
Outline of the Report, 16

2 BACKGROUND OF THE NATIONAL CLIMATE ASSESSMENT   19
Goals of the NCA, 19
Process Used to Create the NCA, 21
Broadening Audiences and Goals, 22
Past Evaluations of the NCA, 25
Findings, Conclusions, and Recommendations, 25

3 FRAMEWORK OF AN EVALUATION   27
Overview of Evaluation, 28
Conclusions and Recommendations, 42

4 NETWORK ANALYSIS AND A NETWORK OF NETWORKS   43
Network Analysis, 45
Some Concepts and Uses of Network Analysis, 46
Applying Network Analysis to Evaluation, 48
Findings, Conclusions, and Recommendations, 51

5 CRITERIA FOR CHOOSING AUDIENCES TO INCLUDE IN THE EVALUATION   53
Audiences Providing Information Needed for an Evaluation, 54
Feasibility of Examining Particular Audiences, 56
Evaluation Staging, 57
Conclusions and Recommendations, 57

6 METHODOLOGIES AND APPLICATIONS TO PARTICULAR AUDIENCES   59
Overview of Potential Methodologies, 59
Illustrative Applications to Particular Audiences, 66
Conclusions and Recommendations, 74

7 IMPLEMENTING THE EVALUATION FOR CONTINUOUS IMPROVEMENT   75
Timing and Sequencing of Evaluation Work, 75
Communication of Evaluation Findings, 77
Legal and Process Considerations, 77
Choosing a Contractor, 78
Conclusions and Recommendations, 79

8 PUTTING IT ALL TOGETHER   81
Asking the Right Questions, 81
Dealing with Multiple Audiences, 83
Choosing Appropriate Methodologies, 84
Continuous Evaluation and Improvement, 85
Multistep Approach to Evaluation, 85

REFERENCES   87

APPENDIX A: COMMITTEE MEMBERS   99
APPENDIX B: NCANET   103
APPENDIX C: CROSSWALK BETWEEN STATEMENT OF TASK AND OVERARCHING EVALUATION QUESTIONS   109
APPENDIX D: NETWORK ANALYSIS: ADDITIONAL MEASURES AND STRATEGIES   113
APPENDIX E: DERIVATIVE PRODUCTS   117
APPENDIX F: FURTHER INFORMATION ABOUT FEDERAL PROGRAMS   119

Boxes, Figures, and Tables

BOXES

S-1 Statement of Task, 2
1-1 Statement of Task, 13
2-1 Public Interest in Climate Change, 23
4-1 Information from the National Climate Assessments Spreads Along Networks, 44
4-2 Example Types of Relevant Connections for Evaluation, 45
4-3 Illustrative Networks That Can Be Studied as Components of a Network of Networks Transmitting Knowledge from the NCA, 49

FIGURES

S-1 Illustrative logic model for an evaluation of the National Climate Assessment (NCA), 4
2-1 Budget crosscut for the U.S. Global Change Research Program (USGCRP), 21
3-1 Attribution and contribution, 31
3-2 Sample logic model, 32
3-3 Illustrative logic model for an evaluation of the National Climate Assessment (NCA), 35
3-4 Sample user pathway through the logic model, 38
4-1 Transmission of information from the National Climate Assessment (NCA), 44
8-1 Illustrative logic model for an evaluation of the National Climate Assessment (NCA), 82

TABLES

2-1 Expansion of National Climate Assessments (NCAs) over Time, 22
3-1 Preliminary Overarching Evaluation Questions for the National Climate Assessment (NCA), 39
6-1 Selected Advantages and Disadvantages of Data Collection Methods, 61
6-2 Data Collection Methods and Considerations for Overarching Evaluation Questions, 64
C-1 Crosswalk Between Themes and Questions from the Committee's Statement of Task and Preliminary Overarching Evaluation Questions Presented in Chapter 3, 110
F-1 USDA Climate Change Hubs Accomplishments During 2013–2023, 121

Key Terms Used in This Report

Audience The receiver or user of climate change information, including those affected by climate change. The Global Change Research Act (GCRA, 15 U.S.C. Chapter 56, Public Law 101-606, 104 Stat. 3096–3104) of 1990 requires the U.S. president and Congress to be audiences of the National Climate Assessment (NCA).

Case study An in-depth study, such as of one group or organization, that is intended not to provide statistically representative results but rather to allow a more thorough investigation than would be possible in a survey.

Cognitive interviews Interviews to determine the thought process that a person goes through to complete a task. They are often used for testing survey questionnaires (e.g., to find out how the respondent would interpret the question, whether the question has the right choices, whether the respondent has the requested data readily available, what the respondent would do to answer the question).

Contextual factors Geographic location and conditions; political, technological, environmental, and social climate; cultures; economic and historical conditions; language, customs, local norms, and practices; timing; and other factors that may influence the outcomes of interest.[1]

Continuous improvement Ongoing learning and evaluation to inform innovation and enhancement of both processes and products over time.

Evaluation A systematic process to determine merit, worth, value, or significance.[2]

Evaluation question High-level questions used to guide an evaluation.[3]

Focus group A qualitative research tool in which a small group of people is collectively asked to respond to typically broad, open-ended questions, allowing interaction among the focus group members. These are useful for understanding what issues are important to people and exploring their motivations and behavior patterns.

---

[1] See https://www.eval.org/About/Guiding-Principles.
[2] See https://www.eval.org/Portals/0/What%20is%20evaluation%20Document.pdf.
[3] See https://www.evalcommunity.com/evaluation-glossary.

Global change Climate and other environmental changes that affect ecosystems, people, places, and societies around the world. Climate change includes variations in average temperature, shifts in precipitation patterns, and intensity of storm events.

Impact evaluation Assesses program effectiveness in achieving its ultimate goals.[4]

Logic model Provides a pictorial theory of change showing the process by which change is expected to happen in complex systems, helping evaluators make hypotheses about the connection between users and the NCA.

Network A collection of nodes and the connections between them; for example, a professional organization might maintain a network through which its members (nodes) receive and distribute information (connections).

Network of networks A multilayer network in which the different layers feature different types of nodes, with potential connections between different types of nodes across the layers.

Nongovernmental organization A nonprofit organization that operates independently of any government.

Outcome/effectiveness evaluation Measurement of program effects in the target population by assessing the progress in the outcomes or outcome objectives that the program is intended to achieve.[5]

Participant One who actively contributes in some way to the development or dissemination of the NCA.

Pathway In a logic model, a pathway reflects the causal path by which an action produces change.

Process/implementation evaluation Determines whether program activities are operating or have been implemented as intended.[6]

Program evaluation A systematic method for collecting, analyzing, and using data to examine the effectiveness and efficiency of programs and to contribute to continuous program improvement.[7]

Stakeholder One who has an interest in or concern about climate changes, including researchers, policymakers, decision makers, other diverse groups, and those affected by climate change; in this report, the terms *audience* and *participant* are used in place of *stakeholder*.

Users People, organizations, agencies, and institutions who make use of climate information to take action. Direct users will include those who use the NCA without an intermediary. Indirect users will include those who use materials created from the NCA without necessarily being aware that the NCA is the source.

---

[4] See https://www.cdc.gov/std/program/pupestd/types%20of%20evaluation.pdf.
[5] See https://www.cdc.gov/std/program/pupestd/types%20of%20evaluation.pdf.
[6] See https://www.cdc.gov/std/program/pupestd/types%20of%20evaluation.pdf.
[7] See https://www.cdc.gov/evaluation/index.htm.

Summary

The National Climate Assessment (NCA) was created in response to the Global Change Research Act (GCRA) of 1990.[1] The GCRA created the U.S. Global Change Research Program (USGCRP, or Program), and called for a scientific assessment that "integrates, evaluates, and interprets the findings," "discusses the scientific uncertainties," "analyzes the effects of global change," and "analyzes current trends in global change." So far, there have been five NCA reports, along with additional reports and materials on specific topics.

The NCA goes through an extensive review process for technical accuracy, including by federal agencies and through multiple opportunities for public comment. These reviews have focused on the technical content of the report, but what has been less studied is the users and uses of the NCA and how it has informed decision-making. An evaluation of the NCA was completed in 2016, which focused largely on the process for creating the NCA report.

Perceiving a need for an updated evaluation with a broader focus, USGCRP asked the National Academies of Sciences, Engineering, and Medicine to convene an expert committee to develop a strategy for examining the uses of the NCA. In response, the National Academies formed a committee of experts in the development process for the NCA, climate communication, the uses of the NCA, and engagement with audiences and participants. Notably, the committee was not charged with conducting an actual evaluation or a research design, but with preparing a strategy for evaluation design. See Box S-1 for the committee's Statement of Task.

Working with USGCRP, the committee first sought to understand the purpose of the evaluation it was charged with informing. USGCRP clarified that the evaluation is intended to support a process of continuous improvement by the Program and that the findings of the evaluation may inform ongoing and future NCAs, as well as other USGCRP products. Thus, the committee observed, an evaluation designed to measure the outcomes of the most recent NCA would be insufficient. USGCRP also needed information on why its products were working or not working with particular audiences. This also suggested to the committee that the original list of questions provided in the Statement of Task may not go far enough. For example, to support continuous improvement, it might also be important to look at the process used to create the NCA and to understand more about its uses, including what aspects have been effective or not to different users. The focus of the Statement of Task on the users of the NCA, their awareness of and engagement with the NCA products, and the usefulness of the information contained in the NCA for decision-making critically informed the committee's evaluation strategy.

The committee also determined that it was premature to create a detailed evaluation design. The Statement of Task requested a strategy, not a design, and much more would need to be known about USGCRP's needs and

---

[1] Global Change Research Act of 1990, 15 U.S.C. Chapter 56, Public Law 101-606, 104 Stat. 3096-3104, §106.

> **BOX S-1**
> **Statement of Task**
>
> At the request of the U.S. Global Change Research Program (USGCRP), the National Academies of Sciences, Engineering, and Medicine (the National Academies) will establish an ad hoc committee to develop a strategy for evaluating stakeholder use of the National Climate Assessment (NCA) and selected other USGCRP products. The committee will develop criteria to prioritize the stakeholder groups that should be involved in such an evaluation, a conceptual and methodological framework and design for an evaluation, and plans for data collection and other information-gathering activities. The evaluation strategy will be designed to help determine to what extent these products meet decision and informational needs of selected stakeholder groups. Diversity, equity, inclusion, and justice (DEIJ) principles will be considered and incorporated in the evaluation strategy. The committee will not provide a technical review of the assessments.
> The evaluation strategy will address the following questions:
>
> *Regarding usefulness*
> - How and to what extent have stakeholders found NCA materials to be useful? What specifically has been useful?
> - Does the selection of topics, regions, sectors, and level of detail (e.g., time frame, spatial resolution, subsector concerns) in the NCA (and USGCRP products it references as well as related USGCRP products) adequately address the stakeholders' needs?
> - What decision or informational needs were well-addressed by the NCA? What decisions can stakeholders make given the level of information provided?
>
> *Regarding decision making, future needs and missing information/details/tools*
> - What future needs are anticipated? What additional types of decisions (if any) do stakeholders anticipate they would revisit given different topics and/or levels of detail? What information would be required to meet needs that USGCRP is not meeting already?
> - What decision or information needs did the stakeholders expect would be met by the NCA but were not?
> - For stakeholders whose decision or informational needs were not met by the NCA and selected USGCRP products, what is the reason? What other products/materials, including other USGCRP and non-USGCRP products, did they use, if any?
>
> *Stakeholder awareness and engagement*
> - How aware or involved were different stakeholder groups in the NCA development process and how did this influence their use of the report? For stakeholders who were not previously aware of the NCA development process, how did they become aware of the NCA?
> - How effectively does the NCA development process engage historically marginalized communities and underrepresented stakeholders?
> - How understandable and navigable is the NCA, including the report documents and findings, and underlying supporting data? Is the NCA information presented in a format that informs decision making?

priorities to create a design. Such an effort would likely require a self-study on the part of USGCRP, perhaps with the assistance of a consultant or contractor, and would require a different type of process and schedule than the Statement of Task provided for. In addition, the creation of an evaluation design will require an iterative process, first prioritizing potential audiences and then developing appropriate strategies for each audience. Instead, this report provides a strategy for creating and implementing an evaluation design.

The committee performed several steps to complete its task. It sought presentations from both past NCA staff and users of the NCA. Working to refine the research questions provided by USGCRP, the committee determined that first a logic model was needed, and developed a preliminary illustrative model, realizing that USGCRP's input is needed to finalize the logic model. The Statement of Task did not request a logic model, but the committee felt that creating one is a key step toward developing an evaluation strategy and design. While considering the audiences served by the NCA, the committee also determined that a structure was needed for examining the roles of the various audiences. Because information from the NCA spreads via existing networks, both formal and informal, to form a dynamic web of actors and organizations, the committee decided that an evaluation study would benefit by treating the audiences as having their own networks, and consider communications across audiences as representing a network of networks. Given the multitude of potential audiences, the committee decided that an evaluation study would be forced to prioritize among them and that audiences would differ in the extent to which data could be collected and analyzed. Thus, the committee decided on criteria for prioritization, based first on USGCRP's information needs and second on the feasibility of engaging with them; these criteria can be used to set priorities among the audiences. The committee also explains in this report how different audiences might require different methodological approaches.

In addition to recommending an approach to evaluation design, this report identifies critical questions and decisions to be addressed by USGCRP, to obtain an evaluation that meets its highest-priority needs for enhancing the accessibility, usability, and appropriateness of the NCA, given its wide range of audiences and users and its goal of informing decision-making.

MULTIPLE AUDIENCES AND NETWORKS USE NCA INFORMATION

As specified in the Statement of Task (Box S-1), multiple audiences, with varied needs, might make use of NCA information. This is important in the context of evaluation because evaluators may need to customize approaches for each audience. In addition, the NCA is produced and functions within a complex environment in which there are multiple sources of information on climate change; these sources comprise networks that may retransmit the information from an NCA (with or without attribution to the NCA), customize or translate the NCA information (perhaps by adding new information such as for specific localities), or be based on sources other than the NCA. This situation complicates the task of an evaluator who wishes to focus specifically on the NCA, as it is difficult to isolate its role and impact. Moreover, the committee believes that it is useful to characterize this environment as a network of networks, with multiple nodes through which information is transmitted and modified. In the context of evaluation, developing an understanding of these nodes—specifically, how information from the NCA is used, modified, and transmitted—is thus important for measuring the outcomes of the NCA and understanding how and why it is used.

Recommendation 5-1: In choosing which groups to study as part of an evaluation, the U.S. Global Change Research Program should seek diversity (including a focus on marginalized populations) similar to that of the participants and audiences with which the Program seeks to engage.

Recommendation 4-1: In designing an evaluation of the National Climate Assessment or related products, the U.S. Global Change Research Program should make use of network analysis as a tool for addressing the evaluation questions related to understanding who key actors are, how information is transmitted across multiple entities, which entities serve as key nodes for disseminating information, and how the network of networks supports that flow of information.

LOGIC MODELS TO CONNECT PROGRAMS AND OUTCOMES

In response to the Statement of Task, the committee developed a conceptual basis for the evaluation, using an established approach in evaluation, namely, the development of a logic model or theory of change for the program(s) to be evaluated. If USGCRP has not already done so, this is a first step in creating an evaluation design. A logic model displays how specified inputs, such as the NCA itself, are intended to produce a range of outcomes. Creating a logic model helps to establish what an evaluation should measure and how the relationships across various elements of the logic model should be modeled or considered.

The committee feels that it is currently premature, based on the information available to the committee, for it to develop a fully refined logic model; such a model would require working directly with USGCRP to determine its goals, priorities, and beliefs about how the NCA is engaging with audiences and informing their decisions. Instead, in Figure S-1, the committee provides an illustrative logic model showing how such a model can be created and used, with the expectation that USGCRP could work with evaluators to modify, refine, or replace this illustrative model. The column "What We Do" lists those products and services produced as part of the NCA; "Who With" lists USGCRP's partners and audiences. The next three columns describe the potential outcomes of the NCA for its audiences.

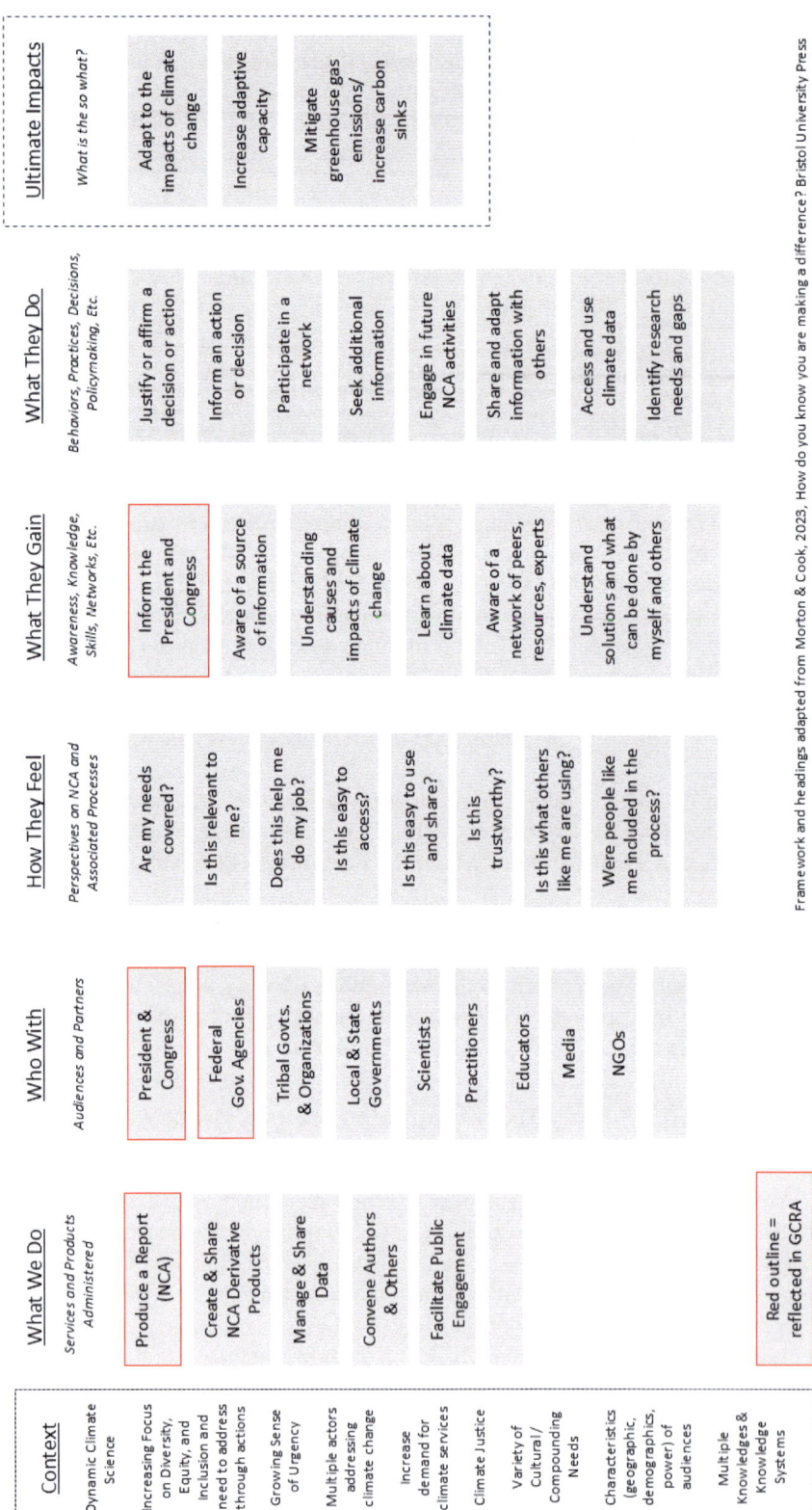

FIGURE S-1 Illustrative logic model for an evaluation of the National Climate Assessment (NCA).
NOTE: GCRA = Global Change Research Act; NGO = nongovernmental organization.
SOURCE: Generated by the committee, adapted from Morton and Cook, 2023.

The column "Ultimate Impacts" lists broader potential societal impacts from the NCA, although these are outside the scope of the planned evaluation effort, which focuses on outcomes for users. Empty boxes are included in the figure to illustrate that these lists may not be complete.

Recommendation 3-3: The U.S. Global Change Research Program should develop a logic model to describe how its products, including the National Climate Assessment, are hypothesized to achieve their intended outcomes.

OVERARCHING EVALUATION QUESTIONS FOR THE NCA

Using the illustrative logic model, the committee developed a list of six overarching questions. These are not intended to be used directly in the evaluation but instead are meant to guide evaluators in developing more specific data collection questions and ultimately measures of use of the NCA and factors affecting its use. The committee recommends that USGCRP and its evaluators build upon these questions by creating their own logic model and derivative evaluation questions.

1. To what extent are priority audiences[2] aware of NCA products and what are the most effective ways to increase awareness? How, if at all, did involvement in the development process contribute to general awareness and use of the report?
2. How, and to what extent, did NCA products address information needs among priority audiences (i.e., what did they gain cognitively in terms of knowledge, skills, attitudes, capacities, etc.)?
3. How, and to what extent, did NCA products address decision needs among priority audiences (i.e., what did they do as a result of using the products)?
4. How did the attributes of the products and process contribute to how users feel, what they gain (e.g., cognition), and what they do (e.g., behaviors)? What about the products and process could be changed to make them more effective?
5. How do the contextual factors described in the logic model influence how audiences feel, what they gain (e.g., cognition), what they do (e.g., behaviors), and how they mediate the use of the NCA?
6. How does the network of networks factor into use?

Note that the sixth question has implications for new types of analysis not included in the previous five. Some types of questions will be relevant regardless of whether one thinks about a network (e.g., what specific parts of the NCA are used, what helps or hinders in accessing or making use of the information), but introducing the concept of a network raises questions such as how the various nodes are connected, which nodes are most used, and what happens to the information as it passes through a node.

Recommendation 3-4: The U.S. Global Change Research Program should use a logic model developed for the evaluation to generate a set of overarching evaluation questions and should consult with partners and selected evaluation users to ascertain whether answering those evaluation questions will meet evaluation users' needs.

PRIORITIZING AMONG MULTIPLE AUDIENCES

The Statement of Task recognizes that the NCA has multiple audiences, and it requests criteria for prioritizing which audiences to include in the evaluation data collection. Both substantive and practical considerations are at play in evaluating outcomes across audiences or selecting which audiences to evaluate. For example, two audiences—Congress and the President—are specifically named in the legislation that created USGCRP and

---

[2] It may not be possible to cover all NCA audiences in an evaluation and thus will be up to USGCRP and the evaluator to prioritize those included in the evaluation. The committee provides a list of criteria for doing so in this report, which are introduced in subsequent paragraphs.

the NCA, and thus may be presumed to be top-priority audiences for the NCA. However, a much wider array of audiences makes use of the NCA or could find it applicable to their decision-making and policymaking. The committee did not attempt to set priorities among these audiences, but instead presents criteria that USGCRP can use to select which audiences to include in the evaluation (see Recommendation 5-1). The criteria should also be weighed alongside available resources: the acceptable financial cost of engaging with a high-priority audience may be greater than for other audiences.

> **Recommendation 5-2: The U.S. Global Change Research Program (USGCRP) should select audiences to include in evaluation based on the following criteria: importance in USGCRP's logic model, including (1) the role of an audience in climate-related decision-making; (2) the role of an audience in the transmission of climate information for decision-making; (3) the generalizability of results from an audience to other populations; and (4) feasibility, diversity, and suitability for the evaluation question and method used. A targeted audience does not need to meet all of these criteria, but the audiences prioritized in an evaluation should meet these criteria collectively.**

TAILORING METHODOLOGICAL APPROACHES TO SPECIFIC AUDIENCES

Depending on the characteristics of an audience and what USGCRP wishes to learn from it, different methodological approaches will apply. In-depth qualitative tools, such as personal interviews and focus groups, can be indispensable for understanding the issues involved, while statistical sample surveys can provide nationally representative quantitative measures of outcomes and barriers.

The size of the NCA audiences evaluated also matters. Some groups, such as K–12 educators or the general public, are so large that evaluators would either need to conduct a survey using statistical sampling (or make use of existing surveys that might have collected relevant data) or choose approaches that are more limited in scope but also less representative (e.g., conducting an online focus group; talking with curriculum developers; working with professional associations, such as the National Science Teachers Association).

For some audiences, it may be best to start with exploratory data collection, such as interviews or focus groups, to learn what types of information people are able to provide and how they interact with climate science information. Audiences that are exposed to climate science primarily through intermediaries may require a different approach than those who deal with the NCA more directly.

> **Recommendation 6-1: The U.S. Global Change Research Program should design its evaluation and data collection plan so that the methods used for priority audiences can answer the overall evaluation questions identified in the logic model. The methods and approach chosen should be tailored to the audience and the evaluation question being investigated.**

USING EVALUATIONS TO SUPPORT CONTINUOUS IMPROVEMENT

Finally, the evaluation of the NCA and its products is best considered not as a "one and done" process, but rather as an iterative process of learning over time. USGCRP made clear to the committee that the evaluation is intended to support a process of continuous improvement and that it was desirable to use the findings of the evaluation to inform ongoing and future NCAs, as well as other USGCRP products. For this, USGCRP needs an evaluation designed not only to measure the outcomes of the most recent NCA, but also to provide information on why its products are working or not working with particular audiences. To support continuous improvement, the committee believes that it might also be important to look at how the process used to create the NCA affects its uses, including what aspects have been effective or not for different users.

At times, exploratory analysis will be needed to determine what data can and should be collected. A case study of an audience (such as a particular nongovernmental organization) may be helpful before conducting a more comprehensive study, and it may provide valuable insights that can help generate lessons learned, to be applied to other entities. If USGCRP creates tracking and feedback mechanisms to be used in the ongoing report preparation and dissemination, such data may be used to develop or improve later data collection. To the extent that the NCA continues to change, there will be value in examining how well those changes support USGCRP's goals.

Recommendation 7-2: The U.S. Global Change Research Program should sequence evaluation into manageable components, allowing for iterative testing and learning about how to best pursue evaluation over time. Sequenced components may include conducting evaluability assessments, piloting focused on certain agencies or chapters of the National Climate Assessment, picking low-hanging fruit first, or developing case studies.

Recommendation 7-3: In communication about evaluation efforts, the U.S. Global Change Research Program (USGCRP) should aim for active two-way communication with users. Communication mechanisms may include ongoing feedback, interim findings, meetings to tailor the communication of evaluation findings to particular situations, and communication about how input was used that helps connect evaluation efforts with USGCRP's objectives.

The size and diversity of the population that does and could use climate information to make decisions means that an evaluation of the outcomes of the NCA and related products is necessarily ambitious. In this report, the committee introduces concepts from evaluation practice, including network analysis, that will be essential to understanding the way the NCA and related products inform decision-making. The approaches recommended will require significant commitment of time and resources from USGCRP. The committee believes that evaluation of use will reveal both the large impact of federal climate science on decisions across the nation and gaps and frailties that can be addressed in the future. Undertaking outcome evaluation is accordingly a major step for the Program. The committee suggests taking this important step to support and improve national assessments to come, and to prioritize USGCRP's efforts on the NCA and related products. At the same time, a targeted effort to understand the outcomes and utilization of the NCA among various audiences can support the more effective allocation of resources by USGCRP and its member agencies, and the understanding of how better to serve the needs of the Program's priority audiences and participants.

EVALUATION LEADERSHIP AND IMPLEMENTATION

To ensure that an evaluation addresses priority needs and opportunities for the Program and the wide range of audiences and decision-makers who use the NCA, engagement and buy-in to the evaluation by USGCRP leadership is critical, as these individuals have the seniority, authority, and responsibility to define the scope. USGCRP must determine the scope of the evaluation; the makeup of the team; the budget; and whether the needs can be met internally, externally, or through some combination of personnel.

Early in the process, USGCRP may consider forming an evaluation team, made of leadership, staff, key partners, and/or others who will guide the evaluation process, and provide diverse perspectives. The committee's recommendations call on USGCRP as the primary actor responsible for decision-making. Additionally, the committee identifies actions and activities to be carried out by evaluators with expertise in designing and implementing evaluation.

Recommendation 3-1: The leadership of the U.S. Global Change Research Program should engage from the start in defining the evaluation scope and should ensure that the leadership perspective, as well as the necessary evaluation expertise, is incorporated throughout the design and implementation of the evaluation.

Recommendation 7-4: The U.S. Global Change Research Program should consider bringing in outside expertise and research capabilities—such as through contractors, consultants, grantees, or interagency agreements—to assist in designing and implementing the evaluation.

1

Introduction

Scientists initially sounded the warning that human activities were changing the global climate in 1957 (Revelle and Suess, 1957). In the seven decades since, climate science has greatly expanded understanding of the scope, pace, and impacts of climate change on human societies and the natural world. It is now possible to obtain detailed projections of changes in climatic conditions and rising sea levels, build models to estimate economic damages and risks, predict altered flow in major river basins, and analyze health risks due to climate-related hazards over the coming three months (USGCRP, 2024), among a host of ways that science can inform decision-making. The scientific basis of this knowledge is validated by peer reviews, tests of consistency, transparency of data, modeling methods, and reporting procedures, contributing over time to a robust consensus. The salience of climate change as a public issue has also risen (Brimicombe, 2022; Tyson and Kennedy, 2023), contributing to a widening range of decisions made possible by the growing scientific understanding of climate change.

Along with scientific consensus, collaboration has increased both nationally and internationally to model and understand the changing climate, collect data on climate changes and their impacts, and develop policies for mitigating and adapting to climate change. Following the creation of the United Nations Framework Convention on Climate Change, the parties to the treaty (Conference of the Parties [COP]) have met regularly; the most recent meeting is known as COP28. At the same time, global climate scientific consensus efforts, most notably culminating in the Intergovernmental Panel on Climate Change (IPCC) Assessment Reports, have produced assessments of the state of the climate, global contributions to climate change, and standardized future policy and emission pathways.

Within the United States, there is a similar need to coordinate and synthesize the growing body of climate science, both to build scientific consensus and to make it accessible to broader audiences. In 1990, Congress enacted and President George H. W. Bush signed into law the Global Change Research Act (GCRA).[1] This statute created the U.S. Global Change Research Program (USGCRP, or the Program), an interagency unit in the White House Office of Science and Technology Policy,[2] coordinating and synthesizing climate research in 15 federal agencies. The GCRA mandated that USGCRP organize an authoritative statement of knowledge of global environmental changes, particularly the changing climate. Such a statement, the National Climate Assessment (NCA), has been released five times, most recently in 2023. The NCA has grown greatly over

---

[1] Global Change Research Act of 1990, 15 U.S.C. Chapter 56, Public Law 101-606, 104 Stat. 3096-3104.

[2] The International Climate Councils Network (ICCN) was launched in 2021 as a forum for climate councils around the world. The United States is not a party to ICCN, but USGCRP functions as a climate council to facilitate coordination and provide data, and the Program works with other nations and their climate councils in various settings.

time, starting at roughly 100 pages and growing to about 2,000 pages, further accompanied by supplemental specialized reports and tools. From the outset, USGCRP has developed the NCA with inclusiveness and transparency as core values. By engaging with sources of knowledge across a wide spectrum and sharing sources, data, and modeling assumptions, this approach can enhance the legitimacy of the findings of climate science among those using the NCA.

Although the understanding of climate change has grown in both the natural and social sciences, there is only a limited grasp of how this knowledge is informing decision-making (see Moss et al., 2019)—the purpose set out in the GCRA. At the request of USGCRP, this committee was created and charged with developing a framework for evaluating the use of the NCA and related products to inform decision-making. To be clear, this committee is not evaluating the NCA but providing advice to USGCRP, and the evaluators it works with, on how to approach the task of evaluating the use of USGCRP products. Designing an evaluation in the current context is complex, with different audiences needing different types of information, and with climate change–related information being disseminated through many sources. Creating an evaluation design specifically to measure the impact of the NCA—outcomes of the NCA in terms of who uses the NCA, how they use it, and how it could be more useful—requires considerable thought. This report, accordingly, aims to inform how USGCRP can prepare evaluations directed at improving future national assessments to make them more useful to the very wide audience of people and organizations affected by and responding to a changing climate.

EVALUATIONS

This section discusses the benefits of evaluations and why this evaluation requires special care.

The Benefits of Evaluations

Program evaluations seek to provide systematic and objective assessments of the effectiveness, efficiency, and impact of programs or interventions (Vedung, 2017). Evaluation provides insights and data to inform decision-making, program improvement, and future planning, as it aims to determine whether and how programs are achieving their intended outcomes. Program evaluation has been used to identify best practices, areas of improvement, and areas where resources could be better allocated; demonstrate accountability to intended beneficiaries, other participants, and funders; and ensure transparency in program implementation.

The use of evaluation in federal climate change–related programs has been modest compared with other fields (e.g., health care), but it can play an important role in helping organizations and policymakers assess programs' effectiveness and impact, including scientific research, collaboration with participants, and providing information to target audiences. Evaluation can also help identify and prioritize gaps in existing research and highlight opportunities to address those gaps to better meet the needs of policymakers and other organizations addressing climate change. Global, scientific, consensus-based efforts, such as the IPCC Assessment Reports,[3] have been the subject of a range of formal and informal evaluation efforts, mostly focused on processes leading up to the production of assessment reports, as well as their effects on climate science (O'Reilly et al., 2024; Schulte-Uebbing et al., 2015; Vasileiadou et al., 2011). These evaluations have helped shape improvements in process, presentation, and communication, contributing to the provision of accessible and widely used reports and other products. The NCA has also been the subject of a number of informal evaluations (Jacobs et al., 2016; Meyer, 2011; Morgan et al., 2005; Moser, 2005; NRC, 2007; Parson et al., 2003) and one formal effort (Dantzker et al., 2016); these have provided suggestions for improving the NCA process and products. As reliable knowledge of a changing climate has become increasingly relevant, continuous improvement in the NCA process is increasingly important. At the same time, a targeted effort to understand outcomes and utilization of the NCA among various audiences can support the more effective allocation of resources by USGCRP and its member agencies, and understanding of how better to serve the needs of the Program's priority audiences and participants.

---

[3] Through USGCRP and its member agencies, the United States contributes to the worldwide research effort that supports the periodic global assessments of the IPCC (2023; see also NRC, 2007).

Why This Evaluation Requires Special Care

Efforts Related to Climate Change Are Widespread

U.S. activities to address climate change are both extensive and diverse, although much of this effort ranges beyond climate science. The president's budget for 2023 proposed over $5 billion for climate science research, and over $18 billion for climate resilience and adaptation programs (OMB, 2022). In a report examining federal funding from 2010 through 2017, the Government Accountability Office (GAO, 2018) identified 18 programs whose primary purpose is to address climate change and 515 additional programs that included other program goals in addition to addressing climate change. For fiscal year 2017, the programs were spread across 19 agencies. These statistics understate the diversity of federally related activities concerning climate change, since, for example, a single grant program may encompass a variety of grants, each with a particular research design and goal. Additionally, climate change–related activities in the United States are not limited to the federal government, but include state, tribal, local, nonprofit, and business activities. To investigate if and how the NCA is informing these broad investments and efforts, an evaluation would need to consider a wide range of actors, the actions they are undertaking, and the climate information that informs their activities.

Actors' Roles Regarding Climate Change Are Highly Variable

Just as many actors are involved in addressing climate change, their roles vary greatly. For example, a federal agency might set national policy, perform or fund research, disseminate information, or carry out activities—such as water management—that rely on climate science information. A nongovernmental organization (NGO) might be involved in disseminating information; it might be taking actions such as planting trees or helping disadvantaged populations; or it might be advocating for policy change. Local city and town officials might be preparing for extreme weather events through better stormwater management or erosion control.

The climate information needs and knowledge levels of these actors vary. They may access different sources of information and their intended uses range from increasing awareness to targeted and specific use of climate projections and data. Their interactions with other actors addressing climate change may range from being highly integrated to being relatively isolated. They may or may not be organized to work together and may be centrally located or highly dispersed.

These variations have implications for how evaluations may be performed. Some groups will be easier to reach than others, and topics that are highly salient to one group may be irrelevant to another. In light of the variety of users and uses of climate information, USGCRP needs to clarify its objectives and the audiences and participants it seeks to reach, both directly and indirectly. This enables an evaluator, working with the Program, to design an evaluation that takes into account the priorities of the Program and provides useful insights into the outcomes of the NCA. Guidance to inform this design is the objective of this report.

Measures of Impact

The purpose of the NCA is to provide information to support the needs of a wide range of decision-makers, needs that might include developing policy, mitigating climate change, or improving resilience on local to national scales. Measuring the impact of providing information for such a wide range of uses is difficult. There is no practical way of measuring changes in knowledge of climate change across the vast array of NCA audiences. Moreover, different users will need different kinds of information, so a test that is appropriate for one group might be inappropriate for another. Perhaps the most straightforward quantitative measure of outcomes would be a measure of the number of citations of the NCA, but people might use NCA information without citing it, or may even use NCA information without knowing that the NCA was the original source. As such, it will be important to define the scope of the audiences broadly, not limiting the study to those who are known users of the NCA. Only by including potential nonusers and indirect users can one measure the extent of use of the NCA, and it may be that the data from nonusers are the most important for determining barriers to use.

Except for bibliographic searches, an evaluation of the uses of the NCA will largely depend on users' perceptions and self-reports. Evaluation measures might address topics such as people's knowledge of the NCA, their ability to access the information they are seeking, and their success in applying the information. Ideally, such measures will capture both direct and indirect use of the NCA (*indirect use* is when NCA information is transmitted and perhaps modified by intermediaries before reaching the ultimate users). The data will be subjective but still informative, more likely to be helpful for program improvement purposes (e.g., in making the NCA more accessible and identifying unreached groups) than for measuring impact. This would be consistent with USGCRP's interest in continuous improvement. Chapter 3 provides further discussion on different aims of evaluation and concludes that, based on the Statement of Task, USGCRP is most interested in an outcome evaluation rather than an impact evaluation.

STATEMENT OF TASK

The process of generating the NCA has been, from the outset, a conversation between the federal government and the scientific community. In that role, USGCRP has both created and facilitated a network of diverse actors and organizations, which in turn has facilitated information exchanges between researchers and decision-makers, and promoted communication that evolves with each iteration of the NCA. This study is intended to support that continuing conversation. USGCRP asked the National Academies of Sciences, Engineering, and Medicine to convene an expert committee to prepare an evaluation strategy for examining the uses of the NCA. The Statement of Task (Box 1-1) for this study is an articulation by USGCRP of important issues that could, if addressed, improve the usefulness of future USGCRP products, as well as illuminate the contributions and challenges of the most recent NCA. The committee sought to respond to each element of the questions listed in the Statement of Task, as spelled out in Chapter 3. We have articulated an approach to evaluating the NCA that aims to capture its dynamism and complexity, as the nation responds to the challenges of climate change.

The committee anticipates that implementing the approach articulated in this report would require working with the Program to help it conduct a self-examination, which is outside of the committee's Statement of Task. Based on the committee's current, limited understanding, USGCRP has developed programming to address many, varied demands for information on climate change, and these have not been prioritized systemically. Over the years, USGCRP has expanded the focus and anticipated audiences of the NCA, so that the original legislative mandate reflects only a small part of what the NCA now delivers.

Following are some topics to be examined by USGCRP that constitute necessary groundwork for creating a detailed evaluation design. Some of these might be addressed before hiring an evaluator, and others might be examined with the help of the evaluator. As discussed later in this report, creating a logic model could be helpful in making the Program's priorities explicit, providing a strong foundation for an evaluation design. In the absence of such groundwork, this report focuses on the Statement of Task's request for a strategy that might be used to develop a logic model and evaluation design.

- What are the important audiences that the NCA is intended to reach? These should be given priority in the evaluation. The original audiences as specified in legislation are Congress and the president, but a current list of audiences as presented by USGCRP includes federal agencies; state, local, and tribal governments; practitioners (including nurses, farmers, adaptation specialists, urban planners, utility managers, engineers, and grant/research funders, and others); NGOs; private and financial sectors; and other decision-makers (such as people buying a home or making investments). The Statement of Task also emphasizes the importance of diversity, equity, inclusion, and justice principles, as well as inclusion of underrepresented groups. These are too many audiences to be adequately covered in a single evaluation. Moreover, these audiences are highly segmented, requiring multiple data collections. Evaluation also may provide information that helps in setting priorities; for that reason alone priorities may change over time.
- How is the NCA meant to fit with other resources on climate change? Is it meant to complement the other sources, filling in gaps left by the other sources; provide a comprehensive summary of all sources; provide a tool on which others can expand; eliminate the need for any other information sources; or some combination of these goals?

> **BOX 1-1**
> **Statement of Task**
>
> At the request of the U.S. Global Change Research Program (USGCRP), the National Academies of Sciences, Engineering, and Medicine (the National Academies) will establish an ad hoc committee to develop a strategy for evaluating stakeholder use of the National Climate Assessment (NCA) and selected other USGCRP products. The committee will develop criteria to prioritize the stakeholder groups that should be involved in such an evaluation, a conceptual and methodological framework and design for an evaluation, and plans for data collection and other information-gathering activities. The evaluation strategy will be designed to help determine to what extent these products meet decision and informational needs of selected stakeholder groups. Diversity, equity, inclusion, and justice (DEIJ) principles will be considered and incorporated in the evaluation strategy. The committee will not provide a technical review of the assessments.
> The evaluation strategy will address the following questions:
>
> *Regarding usefulness*
> - How and to what extent have stakeholders found NCA materials to be useful? What specifically has been useful?
> - Does the selection of topics, regions, sectors, and level of detail (e.g., time frame, spatial resolution, subsector concerns) in the NCA (and USGCRP products it references as well as related USGCRP products) adequately address the stakeholders' needs?
> - What decision or informational needs were well-addressed by the NCA? What decisions can stakeholders make given the level of information provided?
>
> *Regarding decision making, future needs and missing information/details/tools*
> - What future needs are anticipated? What additional types of decisions (if any) do stakeholders anticipate they would revisit given different topics and/or levels of detail? What information would be required to meet needs that USGCRP is not meeting already?
> - What decision or information needs did the stakeholders expect would be met by the NCA but were not?
> - For stakeholders whose decision or informational needs were not met by the NCA and selected USGCRP products, what is the reason? What other products/materials, including other USGCRP and non-USGCRP products, did they use, if any?
>
> *Stakeholder awareness and engagement*
> - How aware or involved were different stakeholder groups in the NCA development process and how did this influence their use of the report? For stakeholders who were not previously aware of the NCA development process, how did they become aware of the NCA?
> - How effectively does the NCA development process engage historically marginalized communities and underrepresented stakeholders?
> - How understandable and navigable is the NCA, including the report documents and findings, and underlying supporting data? Is the NCA information presented in a format that informs decision making?

- Is the NCA intended to counteract misinformation about climate change?
- What level of training/sophistication is expected on the part of the user? Is the NCA meant to be appropriate across multiple levels of sophistication?
- Can the transmission of NCA information through networks be considered a kind of informal partnership, extending the reach of the NCA?

Answering these questions will help support the evaluation design in many ways, including helping to establish which audiences to examine, indicating the extent to which data on other sources are needed, determining what kinds of data are needed and what questions to ask audiences, and prioritizing audiences to determine the most appropriate methodologies.

Given these ambiguities and unanswered questions, this report does not provide a finalized evaluation design. Instead, consistent with the committee's charge, it provides a framework (or strategy, as called for in the Statement of Task) that can be used by USGCRP to work with its evaluator in creating an evaluation design. In addition (as discussed in Chapter 7), different evaluation techniques will have very different costs and timelines. It is not feasible to develop an evaluation design without understanding the available resources that will be allocated to the evaluation.

This report focuses particularly on a tool often used in evaluation, the logic model: a systematic account of how an organization, program, or project seeks to affect its audiences and the decisions they make. Although not specifically requested in the Statement of Task, the committee feels that a logic model can be a key component in developing and interpreting the evaluation. The logic model spells out the intentions guiding the program and forms the basis of an evaluation. An important stage of the proposed strategy, then, is for USGCRP, working with its evaluator, to develop a logic model, followed by the development of a set of questions to be investigated in an evaluation. Guided by the illustrative logic model it developed, the committee proposes—again, only as an example—a set of overarching evaluation questions, which guide the methods and examples developed in succeeding chapters. Appendix C spells out the relationship between the evaluation questions in this illustration and the questions posed in the Statement of Task.

THE COMMITTEE'S APPROACH

Several challenges have important implications for how an evaluation can be conducted and what information it might provide:

- The NCA's primary purpose is to inform decision-making, not to conduct original research or to implement public policy. Thus, its outcomes are measured by others' use of its information. Much of that information is obtained indirectly, as discussed in Chapter 4.
- The NCA has many actual and potential audiences.[4] Two of these audiences—Congress and the president—are specifically named in the legislation creating USGCRP, but the range of those needing and making use of information on climate science is much larger. The profusion of audiences makes it difficult to develop a comprehensive list. It is also challenging to identify organizations and individuals as are potential but not current users of the NCA; however, their feedback could be important in measuring why some are not using the NCA (e.g., because they view it as irrelevant or inaccessible or are not aware that there is climate science that could be useful to them).
- The NCA is operating in a complex environment with many sources of information for decision-makers and the public, some making use of the NCA and others using independent sources. In such an environment, it can be difficult to isolate the impact of the NCA, with some users (especially those who get NCA information indirectly) unaware that NCA is the source of the information they use. For example, both the news media and professional organizations may publish information from the NCA without audiences knowing the source; those receiving the information may attribute the information to the news media or professional organizations.
- Particularly for the general public, current events can have a large impact on people's attitudes and information, including mentions of climate change in political discourse and the occurrence of extreme weather events. These can be important drivers of public policy, and they form an important part of the context in which USGCRP operates.
- USGCRP expressed interest in using an evaluation to make program improvements, and in making evaluation a continuous activity rather than a one-time event. These goals have implications for what types of data need to be collected and the sequencing of data collection.

Given these complexities, the committee determined that a helpful way to study the spread and use of information by multiple audiences is to think of them as a network of networks (Castells, 1996), as discussed in

---

[4] The Statement of Task uses the term *stakeholders* for these audiences. The committee interpreted this word to refer to audiences and participants with which USGCRP works and seeks to share knowledge. Although commonly used as a synonym, to some audiences the word carries connotations that conflict with the inclusiveness to which USGCRP is committed (CDC, 2022; Reed et al., 2024). The committee prefers to use the term *audience* in this report, except where *stakeholder* appears in a direct quote. Other words whose ordinary language meaning overlaps with *stakeholder* are also used in context, including *participant*, *user*, *contributor*, and *partner*.

Chapter 4. That is, there are multiple individual networks, such as urban planners or health professionals, that exist across levels of government, NGOs, media, and individuals. These networks communicate information, plans, and strategies to their members. USGCRP both intersects with these networks, such as when they distribute information from the NCA, and seeks to increase its collaboration with them by involving multiple audiences in writing and reviewing the report. Through these networks, the NCA may reach some audiences indirectly without being recognized as a source, such as when a professional organization customizes NCA information to serve its members. Viewing the various audiences as a network of networks helps to frame the study design and provide a method for analyzing the data.

Networks and audiences vary greatly in their characteristics, and thus in the feasibility of identifying and engaging them in an evaluation. Some audiences are relatively well-defined and identifiable, such as members of Congress or their staff, or climate programs within federal agencies; others are more difficult to define; still others, such as those not making any use of the NCA, may not be readily identifiable in any systematic manner. For some audiences, case studies may provide instructive data about how audiences interact with climate science information, which could lead to important insights on how USGCRP products could be improved to increase effectiveness. Such studies might also identify unanticipated uses of climate information that require further exploration. Chapter 6 discusses methodologies for evaluating how this might be done with different audiences and the uses that they make of information from the NCA. In sum, audiences of potential users of the NCA may need to be prioritized based on their importance to USGCRP's goals and on their accessibility, and often will need to be approached through multiple stages of evaluation as information is gathered.

The committee discussed at the outset what information it needed to complete its task. Collectively, the committee members brought experience in the development process for NCA; communications, dissemination, and audience engagement; the design and implementation of evaluations; and the use of climate assessments. The committee decided it needed more information on how the previous NCA evaluation was performed, on federal perspectives on the NCA, on NGOs' use of NCA, and on historically underrepresented groups and the NCA. To obtain such information, the committee heard presentations from and interacted with the following speakers in public sessions:

- Mike Kuperberg, Executive Director, USGCRP
- Heather Dantzker, Dantzker Consulting, LLC
- Allison Crimmins, USGCRP, Director, Fifth National Assessment
- Allyza Lustig, USGCRP, member of the NCA staff
- Darrell Winner, Environment Protection Agency, member of the NCA steering committee
- Dan Barrie, National Oceanic and Atmospheric Administration, member of the NCA steering committee
- Juan Declet-Barreto, Union of Concerned Scientists
- Ann Marie Chischilly, Northern Arizona University, Institute for Tribal Environmental Professionals

To address the Statement of Task, the committee conducted both meetings that were open to the public and closed meetings consisting of committee and National Academies staff only. The first meeting, which was closed, was devoted to discussing both the Statement of Task and the composition and balance of the committee. The second meeting, also closed, examined what is already known about the outcomes of NCA, who the audiences are, what the committee needs to know about the various audiences' needs, the organization of the final report, and the priorities for information gathering. The third meeting was hybrid, conducted primarily in person but with some virtual participants, and it had a mixture of open and closed segments. The open part of the meeting was devoted to information gathering, with presentations on the previous evaluation of the NCA, federal perspectives on the NCA, and audience perspectives among historically underrepresented groups. The closed portion was devoted to discussions of the preceding presentations and to organizing writing teams and the structure of the report. Between the third and fourth meetings, subgroups of the committee met in closed sessions to work on each chapter. Each committee member participated in multiple subgroups, and all subgroup decisions and processes were shared with the full committee for feedback and to support coordination across the chapters. The fourth, fifth, and sixth meetings were closed, consisting of reviewing drafts of the report, with the sixth meeting devoted to discussing comments received from the external reviewers.

OUTLINE OF THE REPORT

In addressing its charge to "[design] a strategy for evaluating stakeholder use of the National Climate Assessment and select other USGCRP products designed to help determine the extent to which these products meet the needs of stakeholders to support decision-making" (see Box 1-1), the committee considered the needs and insights provided by USGCRP and other experts in its open session, along with the aims expressed in the Statement of Task. The Statement of Task focuses on the users of the NCA, their awareness of and engagement with the NCA products, and the usefulness of the information contained in the NCA for decision-making; this focus critically informed the committee's recommended evaluation strategy.

In addition to recommending an approach to evaluation design, this report identifies critical questions and decisions to be addressed by USGCRP, in order to obtain an evaluation that meets its highest-priority needs for enhancing the accessibility, usability, and appropriateness of the NCA, given its wide range of audiences and users and its goal of informing decision-making. This focus on enabling decision-making was central to the committee's work, as making information available is not the same as enabling decisions. Decision-making requires useful information, including at the appropriate spatial and temporal scales, in forms that are accessible, usable, and understandable to users; decisions also have important social dimensions that depend on networks of relationships, using shared information to address shared challenges.

In the following chapters, this report provides the committee's approach, aims, and recommendations for a strategy for evaluating the NCA to glean insights and guidance on the users and uses of the current NCA, and to inform development and use of future assessments and other USGCRP products, as part of a cycle of continuous improvement.

Chapter 2 provides the overall context for the evaluation, describing the goals of the NCA, how it was created, how it operates, what audiences it serves, how it has grown and evolved over time, and what previous evaluations of NCA have determined. Each NCA is a statement emerging from an ongoing process. The history of the NCA and USCGRP matters: each NCA is an assessment of the current state of climate change, interpreted in light of the evolving priorities of USGCRP and the wide array of participants enlisted by the Program in assembling the NCA. One notable component of that history, and a fundamental theme in this evaluation strategy, is that the audiences making use of NCA have multiplied. Any evaluation of NCA must consider its expanded mission and consider whether and how it is reaching wider audiences, and whether those audiences find it useful. Furthermore, given that the needs of audiences continuously change, a process of ongoing evaluation may be most appropriate to support continuous improvement in the delivery of timely and relevant information to users.

Chapter 3 discusses the goals of the evaluation in terms of what types of questions should be answered, presenting an illustrative logic model for identifying the most important factors and how they are intended to produce change, what interconnections appear among these factors, and what outcomes are intended and among what populations. For example, what knowledge is the Program trying to share? With whom does it seek to work, both in designing and assembling the NCA and in sharing the assessment? How does the Program believe that users respond to the information in the assessment—do they find it salient, accessible, and useful in practice for making decisions? The model offered here is meant to be suggestive but not prescriptive—first, because USGCRP may develop a different logic model, and second, because the data collection and other inputs may suggest other important lines of analysis that were not anticipated when the logic model was initially designed. Based on the logic model, Chapter 3 examines what might be learned from an evaluation, and suggests key overarching questions that, when reviewed and revised in light of a refined logic model, could be the basis for a planned evaluation. As these questions were developed based on the illustrative logic model, they have a somewhat different structure than the questions in the Statement of Task; a crosswalk between the two sets of questions is provided in Appendix C.

Chapter 4 discusses how network analysis can be used to advance the study. As discussed earlier in this chapter, one can picture the audiences for climate science as being diverse in their orientations and needs, with many different research and policy concerns. Collectively, they can be thought of as a network of networks. The science of networks is an evolving branch of applied mathematics that looks at the connections between nodes and how the nodes affect each other. In this way, network analysis can provide a framework in which a wide range of evaluation methods—including, among others, citations of scientific literature, internet-based queries, focus groups, case studies, and surveys—can be brought together to illuminate the networks through which the climate science in the NCA is used, providing a framework for their work and use of the NCA. Understanding networks is also necessary for understanding the full impact of the NCA since networks often act as intermediaries in disseminating and customizing NCA information to the needs of the various audiences. Determining how networks operate and how they shape the knowledge shared through them may be important in measuring the impacts of the NCA.

Chapter 5 builds on the earlier discussion of diverse audiences for USGCRP products in order to articulate criteria for determining which audiences should be prioritized. Given that USGCRP is unlikely to have the knowledge or resources required to examine all potential audiences of the NCA, the Program will need to prioritize those audiences for evaluation, based either on their importance for policymaking or on their ability to provide useful information.

After identifying priority audiences, an evaluator must determine how to collect data from or about those audiences. Chapter 6 provides guidance for selecting appropriate methods to do so, as well as illustrative examples of how a particular audience might be examined using a suite of methods tailored for that audience. It also illustrates how the overarching evaluation questions might be translated into granular questions that could be used in data collection instruments (i.e., overarching evaluation questions spell out what an evaluation as a whole seeks to learn, while much greater specificity might appear on a survey or interview guide).

Chapter 7 discusses some of the practical considerations that appear when designing and implementing an evaluation and discusses the use of evaluation results to guide improvements in the NCA going forward. Chapter 8 brings together the committee's conclusions and recommendations, as developed in the preceding chapters, in a succinct strategy that USGCRP may follow in designing and implementing an evaluation of the NCA.

Evaluating the NCA is complicated because of the many audiences that the assessment can serve in informing decision-making. The guidance developed in this report will require substantial effort and resources from USGCRP, commensurate with its influence and the need for authoritative and reliable climate information as the nation navigates the impacts of a changing climate. Investing in the evaluation of these critical products may hold substantial benefits. The concepts discussed in the chapters that follow can increase the understanding of how the NCA is used, provide information needed to make future assessments more useful, and aid USGCRP in prioritizing assessment-related activities. The committee is grateful that USGCRP recognizes the need for evaluation by requesting this report, and the recommendations advanced below seek to respond to that need.

2

Background of the National Climate Assessment

The history of the U.S. Global Change Research Program (USGCRP) provides important context for the evaluation of the National Climate Assessment (NCA), with several themes becoming apparent. The number and types of audiences for the scientific knowledge of the NCA have increased, affecting the scope of evaluation and creating difficulties in identifying and contacting each audience. Given the Program's ambition of responding to rapid changes in the scientific understanding of climate change, there are now multiple products to consider when evaluating impact.

This chapter describes the legislative and decision support goals for the NCA, the evolution of the process and context across the five NCAs produced to date, the expanding scope of NCA participants and audiences in the assessment process, and past evaluations of the NCA.

GOALS OF THE NCA

USGCRP's goals have expanded over time. The legislation defines specific goals in terms of supporting the president and Congress, but the need of multiple audiences for climate information have led to a much broader definition of what the NCA is expected to accomplish.

Legislative Goals

The Global Change Research Act (GCRA) of 1990[1] established the USGCRP to "improve understanding of global change" through interagency research coordination. The bill was passed unanimously by the Senate, indicating the importance that Congress placed on developing an understanding of global change; in practice, the focus has been on climate change.

The GCRA calls for a research plan that will "combine and interpret data from various sources to produce information readily usable by policymakers attempting to formulate effective strategies for preventing, mitigating, and adapting to the effects of global change" (GCRA § 203, 15 U.S.C. § 2952). It further requires that the Program "consult with actual and potential users of the results of the Program to ensure that such results are useful in developing national and international policy responses to global change" (GCRA § 102, 15 U.S.C. § 2932).

---

[1] Global Change Research Act of 1990, 15 U.S.C. Chapter 56, Public Law 101-606, 104 Stat. 3096-3104.

Among other requirements, the GCRA mandates the Program to periodically (not less frequently than every 4 years) submit to the president and Congress an assessment regarding the findings of the Program and associated uncertainties, the effects of global change, and current and major long-term trends in global change. The GCRA specifically calls for USGCRP to conduct a scientific assessment that

> 1) Integrates, evaluates, and interprets the findings of the Program and discusses the scientific uncertainties associated with such findings; 2) analyzes the effects of global change on the natural environment, agriculture, energy production and use, land and water resources, transportation, human health and welfare, human social systems, and biological diversity; and 3) analyzes current trends in global change, both human inducted and natural, and projects major trends for the subsequent 25 to 100 years. (Section 106)

Decision Support Goals

The GCRA also specifies that USGCRP provide policy-relevant information to support decision-making. This goal is described in the most recent NCA (2023), focused on "continuously advancing an inclusive, diverse, and sustained process for assessing and communicating scientific knowledge on the impacts, risks, and vulnerabilities associated with a changing global climate" (USGCRP, 2023e).

The need to consider the effects of climate change for vulnerable populations, disadvantaged communities, and Indigenous peoples/communities has been recognized since the early days of USGCRP. Over time there has also been an important societal increase in comprehension of and emphasis on climate justice, which has been reflected at the federal level (EOP, 2021). One way that the NCA and other efforts have addressed the need for inclusive perspectives is by seeking methods for better integrating multiple kinds of knowledge and knowledge systems (USGCRP, 2023c; Orlove et al., 2023; Whyte, 2017). The first NCA (NAST, 2000) included a chapter on Indigenous peoples that has been maintained through all five iterations of the report. The first report also named the need for audience engagement to understand needs for decision-relevant information, and indicated as one of five research recommendations the need to better understand social systems and the factors that determine vulnerability. In 2014, a USGCRP Social Science Task Force published a white paper and journal article with recommendations on integrating social sciences to facilitate the use of climate information to inform action (Weaver et al., 2014a,b). That led to the formation of an interagency Social Sciences Coordinating Committee that has developed research and recommendations on inclusion of social sciences in the NCA and USGCRP products. The fifth NCA (2023c) introduced the chapter, "Social Systems and Justice," by saying that social systems are "responsible for the inequitable distribution of both the benefits of energy consumption and the impacts of climate change" (Marino et al., 2023).

In addition to the assessment processes, it is important to note that federal provision of climate services has been developing and expanding over time (see Figure 2-1). The growing sense of urgency around climate solutions (Leiserowitz et al., 2023; Pew Research Center, 2020) has led to a wide range of actors taking steps to address climate change, including subnational governments, civil society, private organizations, individuals, and others (Petzold et al., 2023). The proliferation of solution seekers has also driven an increased demand for climate services (Tart et al., 2020) and greater focus on spatially explicit climate information to serve local needs (NOAA, 2024b). In the past 2 years, the White House Office of Science and Technology Policy (OSTP) convened the Fast Track Action Committee on Climate Services to develop recommendations about how to coordinate climate services across the federal government. The resulting report (NSTC, 2023) included the overarching recommendation for USGCRP to "expand its research coordination role to provide national leadership in coordination and strategic planning of climate services" (p. 2). Such services are defined as "scientifically-based, usable information and products that enhance knowledge and understanding about the impacts of climate change on potential decisions and actions" (OSTP and FEMA, 2021, p. 6). In practice, what constitutes a *climate service* can take many forms, including information about projected physical, environmental, and social changes and the types of response options that could be adopted. (For an example of the way one federal agency is providing such climate services, see the discussion of the U.S. Department of Agriculture hubs in Chapter 5.) It is important to note that because the NCA and other USGCRP products are themselves climate services, their goals and purpose may change as the federal climate services landscape evolves.

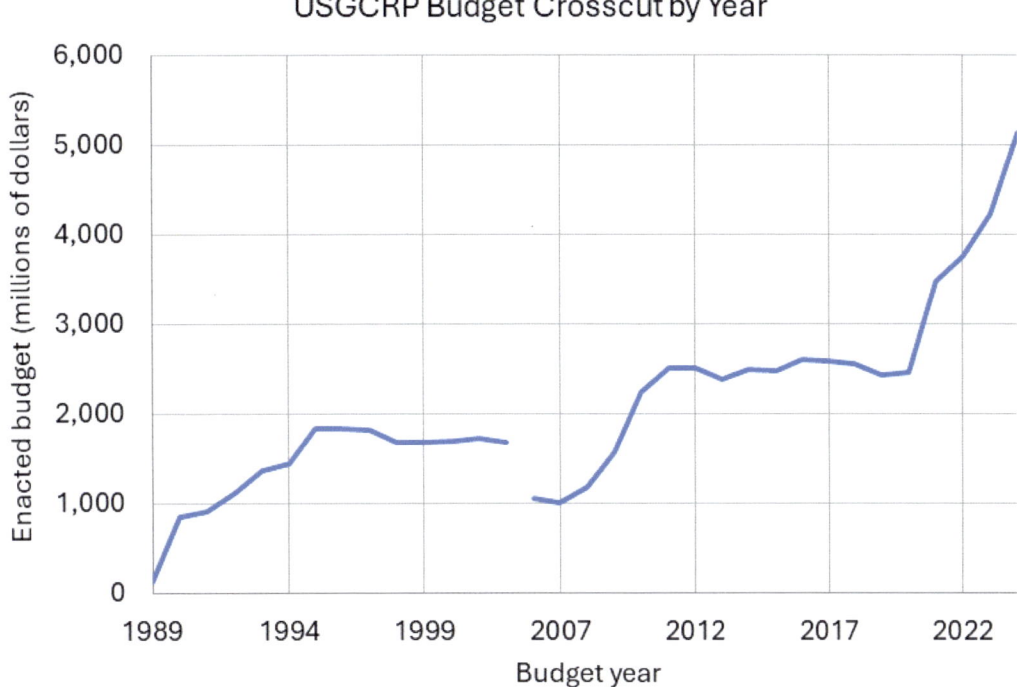

FIGURE 2-1 Budget crosscut for the U.S. Global Change Research Program (USGCRP).
NOTES: The budget crosscut represents the funds identified by USGCRP agencies as their expenditures in support of US-GCRP research activities (USGCRP, 2024). Numbers are taken from the federal budget bills enacted by Congress, in millions, demonstrating the expansion of the Program's work over time. Enacted numbers were not available for 2003–2005.
SOURCE: Generated by the committee based on the budget crosscut reported each year in *Our Changing Planet*, USGCRP's annual report to Congress.

PROCESS USED TO CREATE THE NCA

Each of the NCA reports (Karl et al., 2009; Melillo et al., 2014; NAST, 2000; USGCRP, 2018, 2023c) have been led by a team composed of federal and nonfederal scientists and experts from academia, nonprofit organizations, and businesses. The process (e.g., Avery et al., 2023; NAST, 2000) includes opportunities for public comment. Over time, the opportunities for public input have included information-gathering workshops, regional meetings, opportunities to submit information for consideration, and opportunities to submit comments at various stages of report development.

In addition to public review, the reports are reviewed for technical accuracy by USGCRP member agencies. An external panel of experts also provides a detailed review of the draft report (e.g., the President's Council of Advisors for Science and Technology reviewed the 2000 report [NASEM, 2023]).

Review editors ensure that all comments are addressed adequately. The preface to the archive of public comments and responses to the third-order draft of the 2023 NCA (USGCRP, 2023b) explained,

> Each chapter was assigned a Review Editor to evaluate author responses to both the NASEM review and public comments, and the revised chapter drafts themselves, to confirm that the chapter writing teams had given due consideration to all review comments prior to submission for final agency review and clearance.
> The member agencies of USGCRP sign off on the final report. (p. 1)

The OSTP has sometimes provided a cover letter (e.g., Melillo et al., 2014, p. iii) that becomes part of the report for the delivery to Congress.

The NCA is not policy prescriptive. The GCRA defined it as a scientific assessment, and the first NCA (NAST, 2000) stated "The Assessment's purpose is to synthesize, evaluate, and report on what we presently know about the potential consequences of climate variability and change for the US" (p. 2).

BROADENING AUDIENCES AND GOALS

Over time, the NCA has increased from a little more than 100 written pages to nearly 2,000 (Table 2-1). Much of that increase is due to expanded science and research (Figure 2-1)—for example, improvements in model performance (IPCC, 2021a) and the narrowing of uncertainty around attribution of events to climate change (IPCC, 2021a; NASEM, 2016). The increased length also reflects heightened public interest in climate solutions (see Box 2-1) (Leiserowitz et al., 2023; Pew Research Center, 2020) and an accompanying desire to provide answers to users. For example, the 2018 (fourth) NCA stated that the key messages in each chapter of the report were intended "to provide answers to specific questions about what is at risk in a particular region or sector and in what way" (USGCRP, 2018, p. 4). This builds on USGCRP's mandate to provide policy-relevant information to support decision-making. The GCRA, "like most science policies, has a broader purpose—it constructs an aspirational link between science and some form of social progress" (Meyer, 2011, p. 48).

As noted, the GCRA mandates the president and Congress as the NCA's audience. From the outset, however, USGCRP has recognized the importance of engaging the scientific community and users in preparing the NCA and associated products. The introduction to the initial assessment stated that

> climate science is developing rapidly and scientists are increasingly able to project some changes at the regional scale, identifying regional vulnerabilities, and assessing potential regional impacts. . . . This Assessment has begun a national process of research, analysis, and dialogue about the coming changes in climate, their impacts, and what Americans can do to adapt to an uncertain and continuously changing climate. (NAST, 2000, "About this Document")

All along, these assessments have drawn upon contributions from scientists, as well as

> public and private decision-makers, resource and environmental managers, and the general public. The stakeholders from different regions and sectors began the Assessment by articulating their concerns in a series of workshops about climate change impacts in the context of the other major issues they face. In the workshops and subsequent consultations, stakeholders identified priority regional and sector concerns, mobilized specialized expertise, identified potential adaptation options, and provided useful information for decision-makers. The Assessment also involved many scientific experts using advanced methods, models, and results. Further, it has stimulated new scientific research in many areas and identified priority needs for further research. (NAST, 2000, p. 2)

Chapters 4–6 discuss this engagement with a diverse population of participants and audiences and its implication for evaluating the NCA.

TABLE 2-1 Expansion of National Climate Assessments (NCAs) over Time

	Pages		
NCA Year	**Long Version**	**Short Version**	**Chapters & Appendixes**
2000	522	163	23
2009	N/A	196	25
2014	841	148	36
2018[a]	1526	196	34
2023	N/A[b]	144	42

[a] 2018 details are for "Volume 2"; Volume 1 was a science report of 477 pages released separately.

[b] The web version of the 2023 report was report of record; printed page counts are not available.

SOURCE: Generated by the committee.

> **BOX 2-1**
> **Public Interest in Climate Change**
>
> Public interest in climate change has increased greatly in recent years, as reflected through the media, dedicated climate communication organizations, and educational resources.
>
> In recent years, the number of news stories that link disaster-causing extreme weather events to climate change drivers has increased (Brimicombe, 2022), both as a result of the increasing frequency and size of major disasters (NCEI, 2024) and because attribution science has matured (IPCC, 2021a; NASEM, 2016). Organizations such as World Weather Attribution, Climate Central, and the American Meteorological Society (Blunden et al., 2023) perform attribution studies of extreme events and summarize their findings in press releases intended for both journalists and the broader public. These and other organizations work to foster literacy among journalists. This may be why explanations of extreme weather in the news media tend to adhere closely to attribution science (Brimicombe, 2022; Clarke et al., 2022).
>
> Public interest is also reflected in the inclusion of climate change in the classroom. More than half of states include climate change in their K–12 science standards (Cho, 2023), although few states provide extensive coverage of the processes involved (Bloom, 2021). Global change is one of nine units in the College Board's (2021) Advanced Placement Environmental Science course, although no national standards exist for teaching about climate change at the K–12 level (MECCE, 2022). Resources for teaching and learning about climate change are available for educators, parents, and students through governmental agencies (e.g., the Environmental Protection Agency) educational institutions (e.g., Scripps), and nongovernmental organizations such as the Nature Conservancy. At the college level, climate change is taught as a component of courses in many disciplines. A compilation identifies 27 master's programs in climate science,[a] typically viewing climate through lenses of energy policy, environmental law, justice, or sustainability. Recently, the State of California established the California Center for Climate Change Education to serve as a training resource for clean energy and green technology jobs at the community college level.
>
> ---
> [a] See https://www.masterstudies.com/masters-degree/climate-studies.

The role of users of the NCA and related products is reflected in a call to include them in the 2009 (second) assessment (Karl et al., 2009):

> A vision for future climate change assessments includes both sustained, extensive stakeholder involvement, and targeted, scientifically rigorous reports that address concerns in a timely fashion. The value of stakeholder involvement includes helping scientists understand what information society wants and needs. In addition, the problem-solving abilities of stakeholders will be essential to designing, initiating, and evaluating mitigation and adaptation strategies and their interactions. The best decisions about these strategies will come when there is widespread understanding of the complex issue of climate change—the science and its many implications for our nation. (p. 158)

Around the same time, the National Research Council released a report that emphasized the need for participation, with a core recommendation summarized as follows:

> Public participation improves the quality of federal agencies' decisions about the environment. Well-managed public involvement also increases the legitimacy of decisions in the eyes of those affected by them, which makes it more likely that the decisions will be implemented effectively. This book recommends that agencies recognize public participation as valuable to their objectives, not just as a formality required by the law.[2]

Like the first two assessments, the third NCA (NCA3) (Melillo et al., 2014) was developed by a federal advisory committee (FAC) as an institutional mechanism for engaging nonfederal scientists and experts in the

---
[2] See https://nap.nationalacademies.org/catalog/12434/public-participation-in-environmental-assessment-and-decision-making.

development process. As part of the effort to build on the recommendation for sustained engagement, the FAC was expanded to 60 members and was supported by a team of 300 authors.

> Stakeholders involved in the development of the assessment included decision-makers from the public and private sectors, resource and environmental managers, researchers, representatives from businesses and non-governmental organizations, and the general public. More than 70 workshops and listening sessions were held, and thousands of public and expert comments on the draft report provided additional input to the process." (Melillo et al., 2014, p. iv)

To support the transparency and the usability of the information, the report's references and data sources were linked in the web version of the report, and a "traceable account" for each key finding explained the evidence and rationale supporting it.

The team for NCA3 sought to develop a process focused on ongoing development of knowledge and incorporation of experience to better support the "ability to understand, predict, assess, and respond to rapid changes in the environment" (Melillo et al., 2014, p. 720). The FAC produced a report (Buizer et al., 2013) with recommendations on increasing engagement in the NCA process to achieve those goals, and NCA3 included a chapter (Hall et al., 2014) on "sustained assessment."

A group called NCANet was also convened by the FAC and hosted by USGCRP to extend the reach of the NCA process and products for NCA3. NCANet was a "network of networks" (see Chapter 4) with participants from private, nonprofit, and public sectors who were interested in assisting their networks (audiences, communities) in contributing to and using the NCA. Membership was free and open to any group that wished to engage with the NCA in an ongoing way. As the process for NCA3 continued, the number of organizations represented in NCANet grew to more than 200 (see Appendix B). Members saw their networks as both audiences and participants in the NCA, and they participated in development of communications plans and other activities to ensure that their communities would be able to use the information. This network was seen as a key step toward sustained engagement with audiences and development of an assessment process that would produce more timely and useful information. "By cultivating a network of collaborators who connected the NCA to other networks, the NCA3 engagement process laid the groundwork for a sustained assessment" (Cloyd et al., 2016, pp. 39-40). Although NCANet succeeded in creating a network of networks, and the effort was continued through the fourth NCA (NCA4), staff support from USGCRP was subsequently withdrawn, and NCANet did not continue.

Following NCA3, from 2016 to 2017, the Department of Commerce sponsored a 15-member FAC (the Advisory Committee for the Sustained National Climate Assessment) on behalf of USGCRP to provide advice on engagement with audiences and a "sustained assessment" process. When the FAC's term ended, the members completed their report and recommendations independently (Moss et al., 2019). That paper provides a useful discussion of the case for broad and sustained involvement by audiences and participants in the NCA process and puts forward an ambitious proposal for a civil society consortium to work with USGCRP. Moss et al. (2019) is a useful companion to this committee's work and describes a concrete way to address the matters discussed in this report.

Around the same time, USGCRP agencies formed the Sustained Assessment Working Group, a federal, interagency group focused on assessment process, strategy, and opportunities. Its stated objective was to "develop a process that includes activities inside and outside the Federal Government, makes efficient use of limited federal resources, and—importantly—is informed by and responsive to evolving user needs" (Avery et al., 2018, p. 1046).

Released in 2018, NCA4 was informed by the prior work and attempted to build on the sustained assessment process. For the first time, the team responsible for development of the report was a federal steering committee composed of agency representatives (as opposed to an FAC with both federal and nonfederal members), and the overall author team grew to more than 500 members. NCA4 also added a number of new chapters in response to public input, including subdividing some regions and adding new topics and contexts. Additionally, NCA4 increased the focus on providing localized information. USGCRP staff and author teams held listening sessions and workshops during the development of the report and engaged audiences through a wide range of presentations.

The fifth NCA, released in 2023, continued to build on the development of the fourth, introducing a chapter on economics, increased foci on social systems and justice, new modes for public engagement (e.g., the Art X Climate initiative), updated figures for improved communication, and further improved the delivery of localized information and accessible data (Avery et al., 2023). The number of authors and contributors again increased, and author teams and staff conducted outreach and engagement during the development of the report. After the report's release, USGCRP staff and author teams engaged audiences via new methods, such as chapter-focused webinars, podcasts, and an interactive atlas.

This trajectory of expanding both participants and audiences in the assessment process poses challenges to evaluating the NCA. Perspectives on these challenges and ways to address them are discussed in Chapters 3–6.

PAST EVALUATIONS OF THE NCA

A number of published papers have delved into evaluation of the NCA (e.g., Jacobs et al., 2016; Meyer, 2011; Morgan et al., 2005; Moser, 2005; NRC, 2007; Parson et al., 2003). All of these papers call for continuous learning and evaluation to deliver timely and relevant information to users. Common themes include the need for ongoing user engagement, ongoing assessment processes, and continuous attention to evaluation in order to inform improvement of processes and products. These themes have also been echoed and expanded on in other publications (Buizer et al., 2013, Moss et al., 2019; NASEM, 2017, NRC, 2008, 2013) in ways that are relevant to evaluation of the NCA and related products. These include recommendations for developing an ongoing assessment process and building continuous audience engagement that is thoughtfully scoped to inform development of more useful climate information.

After NCA3 (2014), USGCRP held a workshop on evaluation frameworks, involving experts from across the nation, as a first step toward incorporating ongoing evaluation of the type envisioned for the "sustained assessment" process. The staff produced a report (USGCRP, 2014) summarizing the workshop discussions and recommendations. The workshop focused largely on the process of conducting the assessment, as the process for NCA3 had been reimagined to increase engagement.

The workshop report formed the basis of a post hoc evaluation of the process for NCA3 that was completed by a contracted team of evaluators led by Dantzker Consulting, LLC, in 2016. The evaluation (Dantzker et al., 2016) identified lessons learned and opportunities for improvement, and provided recommendations that can be applied to future NCA planning, development, engagement, and outreach efforts. These efforts were supported by staff from USGCRP and the NCA Technical Support Unit (of the National Oceanic and Atmospheric Administration via North Carolina State University's North Carolina Institute for Climate Studies). The evaluation addressed questions about the structure and inclusivity of the assessment process, resources and materials available to authors, the writing and review process, communications among the groups involved, the engagement process, distribution of products, perception of products, and the ways in which the assessment informed climate-related decisions.

FINDINGS, CONCLUSIONS, AND RECOMMENDATIONS

Finding 2-1: Since the first NCA, USGCRP has drawn on external contributors, as well as federal agency staff, who have participated in the development and writing of the assessments and related products. The number and variety of these contributors has generally increased with each succeeding NCA.

Finding 2-2: Evaluation has not been a regular aspect of the development process for the NCA or other USGCRP products, nor has it been used to evaluate the outcomes of these products.

Finding 2-3: While the usefulness of a network-of-networks approach to extend the reach of the NCA process and products has been recognized previously, it is not a formal part of USGCRP's work at present.

Conclusion 2-1: Themes that emerged from the year 2000 to the present are relevant to the evaluation of the NCA and related products; themes include recommendations for an ongoing assessment process, ongoing audience engagement, careful attention to the scope and purpose of audience engagement, and continuous learning and evaluation to deliver timely and relevant information to users.

Conclusion 2-2: USGCRP works within and has responded to a dynamic context that includes dramatic advances in climate science and dramatic increases in the salience of climate change among the public, eliciting responses in communities, businesses, and subnational governments. Other major changes include a greater focus on spatial variations in the impacts of climate change, as well as increasing attention to the diversity and equity implications of climate change and climate services.

Recommendation 2-1: The U.S. Global Change Research Program, in designing any evaluation of the National Climate Assessment and its other products, should take into account the diversity of participants and audiences with which it seeks to engage.
(The committee further develops Recommendation 2-1 in Chapters 4 and 5.)

Recommendation 2-2: The U.S. Global Change Research Program, in considering evaluation for the National Climate Assessment and its other products, should plan on a strategy for evaluation that allows ongoing learning about how the processes and products are informing decisions, in order to support continuous improvement to its processes and resulting products.
(The committee further discusses Recommendation 2-2 throughout report, including in Chapter 7.)

3

Framework of an Evaluation

Two central goals of the National Climate Assessment (NCA) are to "combine and interpret data from various sources to produce information readily usable by policymakers attempting to formulate effective strategies for preventing, mitigating, and adapting to the effects of global change" (Global Change Research Act of 1990 [GCRA] § 104(d)(3), 15 U.S.C. § 2934(d)(3)) and to "consult with actual and potential users of the results of the Program to ensure that such results are useful" (GCRA § 102(e)(6), 15 U.S.C. § 2932(e)(6)). Any evaluation of the NCA or other products of the U.S. Global Change Research Program (USGCRP) needs to acknowledge these ambitious goals. The committee sets out a way to do that in the next four chapters.

This chapter begins with an overview of key evaluation principles, including defining and identifying different types of evaluation. It includes a description of core components of common evaluation frameworks and applies them to the task at hand. One of the foundational elements of an evaluation is developing a preliminary understanding of the ways in which the program being evaluated might achieve its goals, often depicted in the form of a logic model. That logic model will then be tested and refined through the course of the evaluation.

In this chapter, the committee presents a preliminary logic model, both to demonstrate its utility in evaluation and to serve as a starting point for evaluators. It is also a starting point for much of the discussion appearing later in this report. Based on the logic model, this chapter concludes with a list of the overarching questions that an evaluation might seek to answer. The logic model and overarching evaluation questions, when finalized by an evaluation team, would lay the groundwork—or conceptual model—for an evaluation.

It is worth noting that creating and using the NCA and related products is a social process, and that an evaluation should inquire into the activities of USGCRP, its federal agency partners, the author team of the NCA, and the wide range of people who make use of the NCA both directly and indirectly. The substance of the NCA and related products draws heavily on the natural sciences, including extensive datasets; models; and tools for utilization, such as the NCA Atlas. An evaluation of the NCA needs to draw upon the social sciences to frame and conduct the evaluative process. Both the natural and social sciences provide benchmarks and insights into the substance of federal climate research and are needed to gauge the value of the information being shared in the NCA.

Because an evaluation of the NCA is a complex inquiry into multiple complex social processes, requisite expertise must be engaged from the beginning and throughout the process. First, USGCRP must determine the scope of the evaluation, what the team will look like, the budget, and whether the needs can be met internally or externally or through some combination of personnel. At this stage, engagement and buy-in to the evaluation by USGCRP leadership is critical as these individuals have the seniority, authority, and responsibility to define the scope. Early in the process, USGCRP may also consider forming an evaluation team, made of leadership, staff, key partners, and/or others to guide the evaluation process and provide diverse perspectives. USGCRP may consider hiring one or more dedicated staff to work directly with the evaluator(s), acting as a liaison between the evaluation activities and the broader organization. In hiring evaluators, USGCRP could consider individuals and teams with experience and skills aligned to the goals of the evaluation, such as familiarity with the program, fit with the organizational culture, and methodological approaches that meet the needs. In summary, there is a lot of flexibility in who engages in the work of the evaluation and how they work with each other to accomplish the goals. USGCRP would benefit from thinking carefully about how to engage its capacity to accomplish the evaluation. As such, the committee's recommendations call on USGCRP as the primary actor responsible for the decisions. However, at times in the text, the committee elaborates on actions and activities that are anticipated to be carried out specifically by evaluators who have expertise in designing and implementing evaluation.

OVERVIEW OF EVALUATION

The American Evaluation Association (AEA, n.d.) defines *evaluation* as "a systematic process to determine merit, worth, value or significance" (p. 1). The U.S. Centers for Disease Control and Prevention (CDC, 2023a) echoes this emphasis on a systematic approach in its definition of *program evaluation*,[1] which it defines as "a systematic method for collecting, analyzing, and using data to examine the effectiveness and efficiency of programs and, as importantly, to contribute to continuous program improvement" (CDC, 2023a, "What Is Program Evaluation"). In the committee's discussions with USGCRP and with prior evaluators of the NCA, the importance of being able to use evaluation findings to enhance future products was clearly articulated. As such, the committee believes that this program evaluation framing responds to both the expressed need and the ongoing development of the NCA.

Types of Evaluation

Following are several types of evaluation, offering the following descriptions (EvalCommunity, n.d.):

- Process/implementation evaluation determines whether program activities have been implemented as intended.
- Outcome/effectiveness evaluation measures program effects in the target population by assessing the progress in the outcomes or outcome objectives that the program is to achieve.
- Impact evaluation assesses program effectiveness in achieving its ultimate goals.

The distinction between the latter two types of evaluations is whether the focus is on more direct, shorter-term goals of a program (in this case, those goals include the ability of USGCRP products to meet the information and decision needs of key audiences, which it accomplishes through efforts such as creating products that are considered relevant and trustworthy and fostering networks of peers) or more distal goals (in this case, those impacts include mitigation and adaptation). This is further illustrated in the logic model that follows.

The prior NCA evaluation, conducted by Dantzker and colleagues (2016) (see Chapter 2 in this report), was primarily focused on understanding the effectiveness of the writing and review process and analyzing the reflections of participants in the process of creating the fourth NCA. While the present report does not emphasize process evaluation, the committee anticipates that USGCRP will continue to collect information on an ongoing basis about what is working well and where there are opportunities for improvement in the processes of developing and disseminating the NCA and other products. For example, USGCRP will likely want to gather feedback on meetings and workshops to be able to implement, test, and adapt process improvements quickly to better understand and meet audiences' and participants' needs (Keith et al., 2017).

The evaluation this committee is suggesting focuses on a set of key audiences and how well the NCA and USGCRP products meet those audiences' decision and informational needs. For this purpose, an outcome

---

[1] The CDC (2023a) defines *program* very broadly, encompassing "any set of related activities undertaken to achieve an intended outcome" ("What Is Program Evaluation"). In this chapter, the committee uses the term *program* to denote the development and dissemination of the NCA and other USGCRP products.

evaluation is appropriate because it will help determine the extent to which the program has the intended effect on prioritized audiences, which is consistent with the committee's Statement of Task.

An impact evaluation would speak to the ultimate impact of the NCA and related products. These impacts include mitigating greenhouse gas emissions, increasing adaptive capacity, and adapting to the impacts of climate change. While an impact evaluation also provides valuable information, measuring such impacts is outside the committee's Statement of Task and is not addressed here. Furthermore, an impact evaluation documenting the impact of the NCA in the complex and multifaceted context of climate change adaptation and mitigation would be challenging (Joseph et al., 2023). Instead, the goal of an outcome evaluation would be to investigate whether USGCRP products are having the intended outcomes of enabling key audiences to make informed decisions. As such, the evaluation would demonstrate a key portion of the theoretical pathway toward achieving those long-term impacts. In that way, it builds understanding of how USGCRP's work contributes to the ultimate policy and societal impacts of adaptation and mitigation.

Although the evaluation approach described in this report is primarily an outcome evaluation, it may include elements of other evaluation types. For example, when the evaluators collect information from those individuals and organizations that participated in the development of the NCA, they will inevitably gather valuable insights on the effectiveness of the process. Similarly, the committee theorizes that the NCA will have a long-term impact on increasing adaptive capacity. When exploring how audiences use the information in the NCA, the evaluation may reveal improvements in those individuals' or organizations' adaptive capacity. This blurriness between boundaries of different evaluation types is to be expected and is desirable from the perspective of supporting USGCRP's goal of using the data to improve future outcomes.

Another important note is that throughout the course of this report, the committee refers to evaluation in the singular. However, this guidance might be applied to multiple evaluations if USGCRP chooses to evaluate different products separately or chooses a continuous process of evaluation and improvement. It is also possible (as discussed further in Chapter 7) that the evaluation may be divided into phases. The approach described here could, therefore, be applied to a single overarching outcome evaluation or multiple evaluations of narrower scope.

Frameworks for Evaluation

The committee reviewed several evaluation frameworks that can be used to design an evaluation focused on impact or outcome. These include the CDC (2023a) Framework for Program Evaluation in Public Health, Utilization-Focused Evaluation (Patton, 2003, 2012), Practical Program Evaluation (Chen, 2005), Participatory Evaluation (Cousins and Whitmore 1998), Contribution Analysis (Mayne, 2012), and Equitable Evaluation (Stern et al., 2019). These frameworks often provide suggested steps for the evaluation process and/or standards for determining whether the evaluation itself is meeting its intended goals. The evaluation team may want to select one particular evaluation framework or use concepts from several to guide their work. This section describes significant components that are common across several evaluation frameworks and can be applied to the evaluation at hand. These components are also highlighted as recommendations for inclusion in the final evaluation.

Describe the program, its goals, and the pathways for achieving those goals. Evaluators often begin by developing a detailed understanding of the program and its intended effects. Clarifying the program and its key goals will help evaluators establish hypotheses for testing how the program is meeting its objectives. Logic models are one common mechanism of depicting the relationships between the program activities and short-term outcomes and longer-term impacts (Julian, 1997). In the next section, the committee presents a preliminary logic model to demonstrate how this can be used to provide direction to the evaluation.

Design the evaluation with its use, intended users, and impacted parties in mind. Utilization-Focused Evaluation is particularly relevant here, as that framework "begins with the premise that evaluations should be judged by their utility and actual use; therefore, evaluators should facilitate the evaluation process and design any evaluation with careful consideration of how everything that is done, from beginning to end, will affect use" (Patton, 2013, p. 1). In order to ensure the usefulness of an evaluation, it is important to consider which are the entities most likely to use the evaluation. This list will overlap with the intended audiences of the NCA and related products (described in Chapter 5), but it is not necessarily the same. For example, USGCRP itself is likely the highest-priority evaluation user, as it will incorporate learnings from the evaluation into future products. Similarly, the funders of USGCRP (i.e., Congress) may use the evaluation to inform decisions about what aspects of the NCA and similar products are most effective and where it is most appropriate to direct resources. The federal agency partners of USGCRP would likely be considered key evaluation users as well. Identifying and engaging these high-priority users in the evaluation design can increase the likelihood that the finished product will meet their needs. This might include holding workshops with key users to develop or refine the logic model, vetting overarching evaluation questions and data collection instruments, and sharing early findings to provide perspectives on interpretations of results and to determine reporting and dissemination strategies.

In addition to users of the NCA, interested and impacted parties also need to be included in the evaluation design. These include potential users who may never read the NCA but have an interest in how it is used.

As described in Chapter 2, supporting continuous improvement has been a priority for USGCRP. As a result, it is important to ensure that the information gleaned from the evaluation will inform efforts to improve the process and products. For example, when designing data collection instruments, the evaluators might want to gather very granular information about what parts of the NCA were most helpful in meeting audiences' decision-making needs and why, as well as which were not helpful and why not. This could lead to specific adjustments in future products.

Engage priority USGCRP audiences in implementing the evaluation. Chapter 5 describes a broad array of audiences for USGCRP products and presents guidance on setting priorities among them. As described in Chapter 6, an evaluation will seek to collect information from a prioritized set of audiences via various data collection efforts (e.g., surveys or focus groups). In addition to collecting information from them, evaluators may seek to engage representatives from those audiences in the design of the logic model, evaluation, and/or interpretation of the results. The extent to which these audiences are engaged in design and interpretation steps can range greatly (Chouinard et al., 2013), but many frameworks encourage some level of engagement. Additionally, attention will be needed on how to sample among and within audiences for the purposes of data collection. Chapter 5 provides guidance on criteria for selecting audiences to include in the evaluation. The evaluator will also need to determine a sampling approach for selecting within prioritized audiences. Because a representative sample may be difficult to obtain given the breadth of audiences, random or convenience samples may not seek to achieve representativeness but instead ensure a diversity of voices is heard.

Design and implement the evaluation with equity at its center. Careful consideration will be needed when designing and implementing the evaluation to place equity at its center. The entire process—the team carrying out the evaluation, the audiences included, the evaluation itself, and strategies to communicate and disseminate its findings—all need to be anchored in equity principles. A growing body of literature emphasizes equity in evaluation design, including work focusing on equity as a key dimension in the evaluation itself, as well as broader, more holistic approaches to an equity-centered evaluation (Kallemeyn, 2009; Stern et al., 2019; Thomas, 2020; Whyte, 1989), which would be appropriate to consult.

Research and practice suggest a number of key principles of equitable evaluation, starting with the evaluation team. Evaluator teams are stronger when they include members from diverse backgrounds, perspectives, and lived experiences. This will help reduce biases and ensure that all voices are heard and considered during the evaluation process. Similar considerations should be applied when engaging interested parties and NCA audiences—evaluators should seek to engage a wide range of individuals from an array of backgrounds, locations, and varying levels of access to resources. When designing the evaluation, evaluators should use methods that are sensitive to the values, norms, and beliefs of the communities or audiences being evaluated, and that respect their time and resources (see Chapter 7). For example, principles from the co-production of knowledge (Djenontin and Meadow, 2018; Jagannathan et al., 2020) or action research (Lewin, 1946), such as reflective practice and action-learning cycles (Smith, 2001), may be appropriate. Evaluation approaches tailored to Indigenous peoples, including principles of cultural humility, use of multiple channels of communication, and learning from or including Indigenous strategic evaluators may also be considered (LaFrance et al., 2012; Whyte, 1989, 2017). Finally, best practices in inclusive communication can be used to reflect back to evaluation participants what evaluators heard and learned, and what is being done with evaluation findings.

Use concepts like contribution analysis to reflect the complexity of USGCRP's products. When looking at complex systems, attribution or cause and effect are difficult to detect and measure because multiple factors may

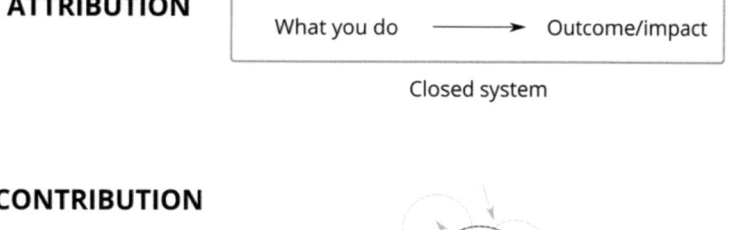

FIGURE 3-1 Attribution and contribution.
NOTES: Understanding cause and effect is important for evaluation. Attribution is easy to document in a closed system but difficult to detect in a complex dynamic system. A contribution model is a better fit for this environment.
SOURCE: Morton and Cook, 2023. © Reproduced with permission of the Licensor through PLSclear.

produce or help produce a single outcome, and with the factors operating in overlapping ways, it is difficult to isolate the effect of each (see Figure 3-1). Contribution analysis (Mayne, 2008) embraces this complexity and fosters understanding about multilayered change in a complex, dynamic system (Montague, 2012; Morton and Cook, 2023). While contribution analysis is still a relatively new concept, it has been applied to evaluations similar to the one proposed here—for example, when assessing the impact of advocacy on policymaking (Kane et al., 2021) or evaluating "large-scale transformational change processes" (Junge et al., 2020, p. 228). It is a multistep process that entails establishing a cause-and-effect relationship to be studied and its associated theory of change; it then uses an iterative process to gather multiple streams of evidence for use in assessing, challenging, revising, and strengthening that contribution story (Mayne, 2011). Core components of this approach include a participatory process for developing the theory of change, the gathering of different types of evidence, and the consideration of alternative causal theories to determine which one is best supported by the evidence (Wimbush et al., 2012).

Logic Models

While many organizations and programs can articulate their actions and desired global impact, elucidating the mechanisms of that transformation proves to be more challenging. An important first step of an evaluation involves establishing an initial conceptualization of how the project, program, or organization to be evaluated attains its objectives, which is typically illustrated through a logic model, which may also be called an outcome map, program theory (Weiss, 1998), theory of action (Patton, 2003) or impact pathway analysis (Morton and Cook, 2023).

A *logic model* is a process-based figure that describes how change is expected to happen in complex systems, setting out the mechanisms that link an activity to a desired outcome or impact (University of Wisconsin Extension, 2024). By making explicit the activities and outcomes, and how they are connected, the logic model spells out the intentions of the program—how it thinks its actions lead to the outcomes desired (CDC, 2018). Logic models are a powerful tool for fostering a shared understanding of the program by the staff, evaluator, and users of the evaluation. The logic model helps the evaluator generate hypotheses about relationships between people, activities, and outcomes that make a program work and how they should be measured. The evaluation then tests these hypotheses against observations.

There are many ways to structure and visualize a logic model, with much variation in format, level of detail, and method of organization (CDC, 2018). In its most basic form, logic models usually describe inputs, outputs, and outcomes or impacts (University of Wisconsin Extension, 2024). However, a logic model may include many other types of information to reflect the nature of the program, including audiences, activities, outcomes at different scales

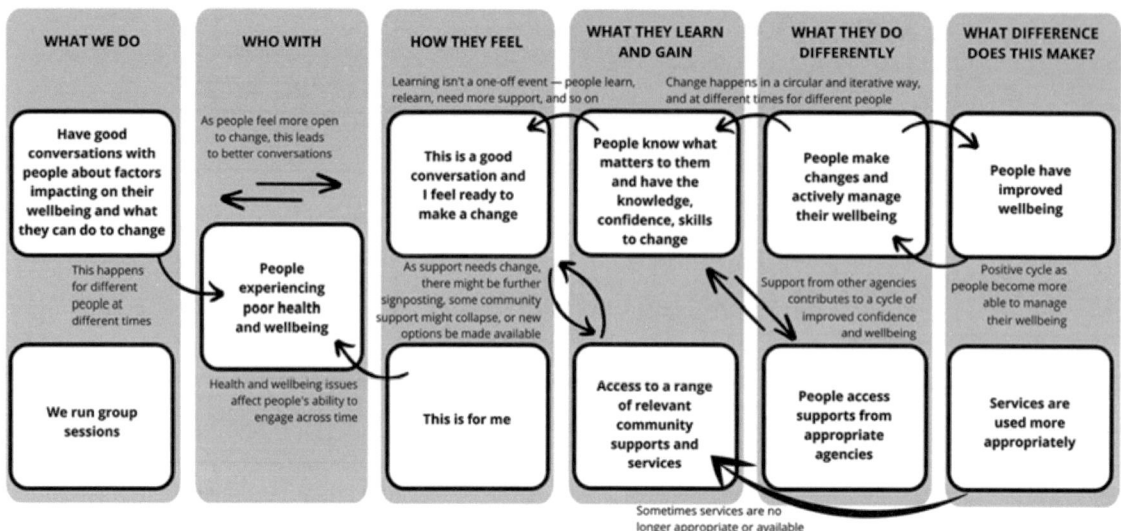

FIGURE 3-2 Sample logic model.
SOURCE: Morton and Cook, 2023. © Reproduced with permission of the Licensor through PLSclear.

(e.g., short term or intermediate), contextual factors, or moderators. Morton and Cook (2023) present an approach to creating a logic model that instead uses short phrases to describe these same basic components (Figure 3-2). If USGCRP chooses to develop a logic model as a foundation for the NCA evaluation, it could consider a variety of approaches to determine which fits best.

To illustrate, this chapter (and report) includes a preliminary logic model for NCA based on the approach presented by Morton and Cook (2023). Their approach assumes that when people gain new perspectives and knowledge, they act in different ways, leading a program to make a difference. The columns (see Figure 3-2) describe key elements of the program. This includes "What We Do," which describes the activities, services, and products generated by the program. "Who With" describes the groups of people who are engaged in the program in various ways. This section could include more than one group of people, and the groups may vary in terms of level of engagement, but anyone served by the program should be identified here. Chapters 4 and 5 address the challenge of identifying and engaging with the wide span of audiences and participants with which USGCRP works.

Outcomes are the changes resulting from an activity or service provided (Blamey and Mackenzie, 2007; Morton, 2015; Rogers, 2008) and are also included in the logic model. Similar, alternative logic model frameworks exist. The Morton and Cook (2023) approach discerns three types of outcomes: "How They Feel," "What They Learn & Gain," and "What They Do Differently." Although observed change in behavior—"What They Do Differently"—is often what is meant by the outcome of a program, Morton and Cook (2023) posit that subjective outcomes matter as well. Outcomes describing "How They Feel" relate to people's feelings, beliefs, and perspectives. Outcomes describing "What They Learn & Gain" relate to changes in awareness, knowledge, or skills. These changes often precede or are associated with a change in behavior. "What They Do Differently" describes the behaviors or decisions that people would be expected to take if the program achieves its intended results. Finally, "What Difference Does It Make?" articulates the broader outcomes the program hopes to achieve. This column presents the vision of change that the program is working toward. In other words, if the program activities lead to the expected outcomes, then the program contributes to making a broader difference in the world.

Morton and Cook's (2023) approach to organizing information in a logic model may be uniquely relevant to NCA in that it encompasses the complex dynamics between attitudes or perspectives, cognitive or rational processes, and decision-making or behavior. For example, dual-process theories explain how at times experiences, emotions, and heuristics guide decision-making, but at other times people process information slowly, deliberately, and analytically (Kahneman, 2011). In the context of program evaluation, how people feel about the information, service, or product being provided is an important outcome to consider because feelings—such as the credibility and legitimacy of the information (Cash et al., 2003) or trust in the messenger (Besley and Dudo, 2022)—precede or are associated with other important outcomes such as attention or use. In addition to feelings about the information, the NCA might prompt feelings about climate change as an issue or it might motivate behavior (Steg, 2023).

Across the breadth of audiences and partners that NCA serves, it can be expected that some will be strongly motivated to attend to this information in great depth. For example, someone working with a climate adaptation science center might read multiple chapters of the report, review the background literature, and think deeply about how to translate the information for a specific set of decision-makers they serve; their attitudinal responses may be less important. In contrast, attitudinal responses to the NCA may be very impactful in determining its use by a local public health official who is just starting to understand the impacts of climate change on human health in their county. Being unfamiliar with climate science or USGCRP, this person may need to feel that the information is trustworthy and useful, before attending to the information in greater depth or sharing it with others.

Although the columns are arranged sequentially in the logic model (see Figure 3-2), they are interrelated, and causality can run in multiple directions. The activities, audiences and partners, and outcomes in a logic model can be connected by specific pathways, as Figure 3-2 illustrates. Pathways specify chains of causation through which a program achieves its outcomes, and they can be organized by audiences, program activities, or impacts (Morton and Cook, 2023). Each portion of the logic model can be used to generate evaluation questions and suggest ways in which the various items within the logic model are interrelated. According to Morton and Cook (2023), the pathways are units of analysis in evaluation and the connections between boxes help define the variables of interest and the types of evidence to be gathered.

While useful for planning, implementing, and evaluating a program, there can be several challenges associated with logic models, their development, and maintenance. A logic model is by nature an oversimplification and may not capture the full range of activities, audiences, and outcomes associated with a program. Furthermore, logic models are based on assumptions; if those assumptions are incorrect or biased, the model may be inaccurate. The process of developing a logic model is time-consuming and usually requires significant input from key partners and audiences. Once created, a logic model ought to be a living document and regularly assessed and updated. Updates are important because the logic model constitutes a theory of change shared by the evaluator and the audiences of the program. An outdated logic model can accordingly become a liability that limits or misinforms the program's response to evolving situations and conditions.

In summary, elucidating the mechanisms by which a project, program, or organization contributes to a desired impact or change in the world is a complicated yet necessary endeavor for conducting an evaluation. A logic model provides a conceptual framework for fostering a shared understanding of how activities link to outcomes. There are many ways to go about developing a logic model; one approach is described here. The model can be developed at the project, program, or organization scale and can be nested to show how all the work of an organization contributes to an anticipated impact. By leveraging logic models as a tool, USGCRP can articulate its intentions and explain the connections that it believes lead to program effectiveness. The logic model may continue to evolve as more nodes are detected and the relationships between nodes are better understood.

Illustrative Logic Model Applied to the NCA

This section presents an illustrative logic model for the NCA, using the approach discussed above. However, because the committee's knowledge is incomplete, the logic model presented here is incomplete. A logic model is usually generated through extensive discussion and deliberation, in which the people who shape and define the program describe who engages with the activities and what they feel, think, or do differently because of their engagement. While the committee offers an example here to aid in the framing of the evaluation recommendations, the report recommends that USGCRP, with evaluators it selects, engage in the process of either refining the logic model presented here or developing a new one that more fully reflects USGCRP's understanding of how its work contributes to informed decision-making.

As shown in Figure 3-3, the illustrative logic model for USGCRP is organized into columns with labels adapted from those in Figure 3-2. "What We Do" includes some of the services, products, and activities administered by USGCRP, including the NCA, climate data, and public engagement activities. This logic model also includes expected outcomes. "How They Feel" includes some of the feelings about NCA that might be related to other outcomes, such as whether people feel it meets their needs or helps them with their job or whether the information is trustworthy. "What They Gain" includes primarily cognitive outcomes, such as awareness of the assessment, knowledge about climate change, or familiarity with sources of climate data. Awareness is particularly key, as it mediates many other outcomes. Awareness of information and knowledge of climate changes and impacts is important but insufficient to prompt action in response.

The next columns in the logic model provide critical insights into conditions and prompts for decisions and actions. "What They Do" includes some of the actions NCA users might take. This includes justifying an existing decision, informing a new decision, or sharing NCA or climate change information with others.

The committee modified the Morton and Cook (2023) model by adding a "Context" section to the logic model, to make transparent the dynamic social environment and background conditions that may affect everything presented in the logic model. (See Chapter 2 for additional discussion of historical and contextual factors.) Contextual factors are important because they may independently cause or alter some of the effects of the NCA (e.g., by increasing or decreasing its perceived credibility, by altering the salience and usability of federal research as a source of information, or by expanding the size and composition of the potential audiences interested in climate change and its impacts on different geographies and populations).

As previously discussed, a logic model also usually includes a description of the broader societal impact, called "Ultimate Impact" in Figure 3-3. In this illustrative logic model, the end goal that the NCA supports is to inform decisions on adaptation to the impacts of climate change, increasing adaptive capacity, and mitigating greenhouse gas emissions. These ultimate impacts unfold over a long period of time, however, in a complex and rapidly changing setting. In a practical evaluation, as called for in the Statement of Task, observing such impacts and relating them to a specific NCA or product is not feasible.

The committee recommends that USGCRP develop a logic model as outlined above (Recommendation 3-3). We highlight several considerations in developing this logic model:

- *Map out the activities and outcomes:* Determine what activities USGCRP carries out in order to produce the outcomes. Define the different types of outcomes and sort them into categories (e.g., "What They Feel").
- *Define the ultimate impact:* Identify the broad impact to which the program contributes. What are the broad and long-term effects that USGCRP and the NCA have on people and society?
- *Develop connections:* Identify high-priority pathways (connections linking boxes). These correspond to the intentions of the Program in creating and disseminating the NCA. These pathways are hypotheses to be tested in an evaluation.
- *Consider the context and assumptions:* Define the contextual factors that are important to program implementation and the outcomes to which implementation contributes.
- *Utilize participatory approaches:* Throughout its history, USGCRP has involved key participants in the preparation of the NCA, both in federal agencies and beyond. It is therefore appropriate to involve some of these participants in developing the logic model to ensure that it reflects important perspectives and priorities.
- *Choose a visual representation:* Use a visual representation that reflects the logic model, so that it is easier to understand.
- *Iterate and revise:* Use the logic model to articulate causal assumptions embodied in USGCRP's intentions. The logic model is a tool for creating a common understanding of the ends and means pursued in the NCA. Thus the logic model needs to be modified as the evaluator prompts USGCRP to reconsider its assumptions in light of the findings of the early stages of the evaluation.

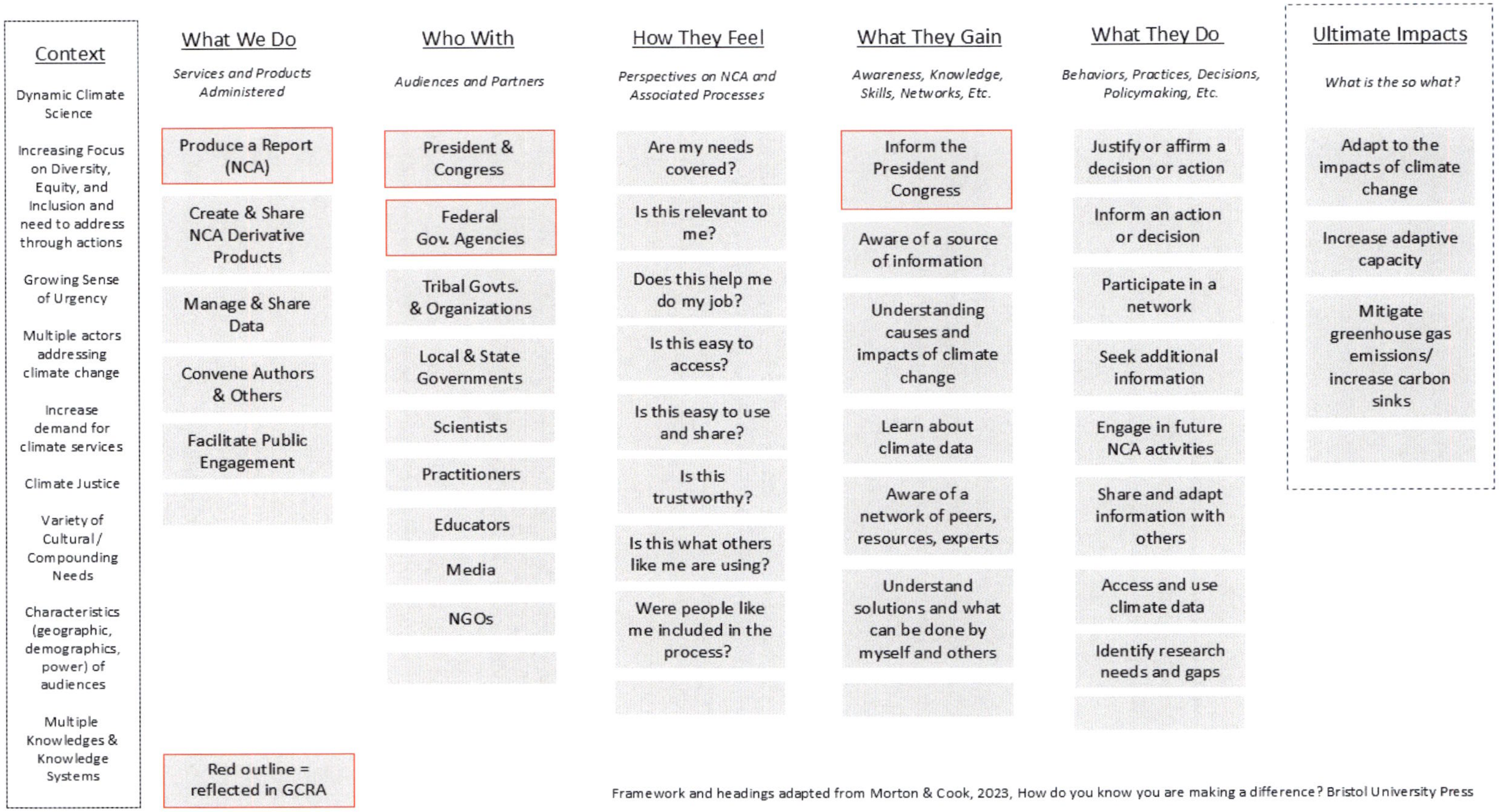

FIGURE 3-3 Illustrative logic model for an evaluation of the National Climate Assessment (NCA).
NOTE: GCRA = Global Change Research Act of 1990; NGO = nongovernmental organization.
SOURCE: Generated by the committee, adapted from Morton and Cook, 2023.

Framework and headings adapted from Morton & Cook, 2023, How do you know you are making a difference? Bristol University Press

Networked Nature of NCA Use

The committee calls specific attention to the outcome labeled "Share and Adapt Information with Others" because USGCRP provides information that is often adapted and shared with others through people and organizations that span boundaries (Goodrich et al., 2020). Many people likely receive extracts, modifications, or customizations of the information from the NCA through other sources rather than from the NCA directly. As described in Chapter 2, USGCRP organized the NCANet in 2012, a formal network that grew to more than 225 organizations collaborating to connect their communities to the NCA process and products. Support for NCANet ended, but outreach to audiences continues. Some audiences who have received climate science information may not know that a National Climate Assessment exists. To these audiences, what matters is that they are getting relevant information from a trusted source—a leader of their network. Ideally, this iterative diffusion process leads to tailored information more amenable to specific uses (Kalafatis et al., 2015), with some organizations even forming strategic partnerships with others in these networks to scale up their ability to translate and engage effectively with information users (Bidwell et al., 2013; Lemos et al., 2014). However, scientific messages can also become distorted as they spread through networks, potentially undermining scientific credibility and resulting in unintended consequences (West and Bergstrom, 2021).

Information diffuses from the NCA within a structure: it spreads via existing networks, both formal and informal. Over time, these networks have become an increasingly complex and dynamic web of interconnected organizations and actors operating from the global to local levels across many sectors. In this sense, dissemination of the NCA and its products can promote social learning throughout these networks. Social learning through networks can in turn affect beliefs and behaviors, and in some cases contribute to collaboration in solving social problems (Bener et al., 2016; Henry, 2020).

Obtaining NCA information through third parties may also result in people using the NCA without knowing the original source. It is important to account for such indirect contacts in order to avoid underestimating the impact of the NCA. Chapter 4 elaborates on this "network of networks" idea, explaining how network analysis can help contextualize this transmission of information. Because of the networked nature of NCA information, an evaluation of use should trace these flows of information, answering questions such as which nodes have the largest role in transmitting information, which are considered most trustworthy, and how the information might be modified or customized in the process.

Defining Audiences, Participants, and Pathways

A key component of a logic model is defining and discussing who the project, program, or organization serves (Bamzai-Dodson et al., 2021). In addition to identifying groups of people the program involves or serves, it is also important to identify those who are not currently being served but who could be reached if specific efforts were made.

Any evaluation will reflect the world views or cultures of those conducting the evaluation and that no evaluation is free of these lenses. The American Evaluation Association (AEA, 2011) published guidance on how to ensure a "culturally competent" evaluation; some of these suggestions also address larger equity issues facing any evaluation. The AEA guidance defines *culturally important factors* as including, but not limited to, "race/ethnicity, religion, social class, language, disability, sexual orientation, age, and gender," as well as consideration of "[c]ontextual dimensions such as geographic region and socioeconomic circumstances" (AEA, 2011, "What is culture?"). A course correction may be necessary if the assessment emphasizes culturally significant factors and the emerging meanings of one group over those of others; an additional concern is that the data produced may not be valid internally and externally for the larger population under consideration (AEA, 2011). Here, the phrase *internal and external validity* broadly refers to whether researchers are measuring what they set out to measure (internal) and whether the results are applicable to broader populations or contexts (external) (Crano et al., 2014). Note that some scholars apply issues of internal and external validity to qualitative designs as well (Thomann and Maggetti, 2020). Fundamentally, questions about validity ask whether the research is both meaningful and trustworthy. If a culturally significant factor is prioritized for use without being a specific goal of the research design, the data may suffer from reduced validity. For example, if the resulting data were intended to measure general usage, and instead all users were from a specific socioeconomic class or geographic area, then the results might not be valid with regard to overall usage; conversely, findings that are valid in general may not be valid for some subgroups. Minorities, who may be disproportionately affected by climate change (or by melioration and mitigation activities), may have needs that are not reflected in the overall findings.

The AEA guidance document specifies that the assessment should strive for *cultural competence*, meaning that the evaluators should be self-reflexive about their own identity(ies) and have the capacity to respect and comprehend other cultural perspectives (AEA, 2011). AEA offers cultural competency practices including (1) recognize the complexity of cultural identity; (2) aim to use methods of knowing and measuring that take into account how different cultures define concepts in ways that are salient to them and may not necessarily align with classical ways of knowing; and (3) be mindful of the evaluators' limitations and worldviews and how these may affect the evaluation (AEA, 2011).

In Figure 3-4, the audiences mandated by the Global Change Research Act[2]—Congress and the president—are identified in red. This is one pathway that can be investigated in an evaluation: does the NCA inform Congress and the president? Figure 3-4 also highlights graphically another, more detailed pathway that depicts how the NCA interacts with federal agencies. Chapters 5 and 6 discuss this and other pathways, in order to illustrate the advice the committee provides on how to design an evaluation. Each pathway identified for study is central to the design of an evaluation because it specifies a hypothesis describing how the NCA contributes to impacts on specific participants and audiences. The evaluation will include gathering information to explore such hypotheses.

Consider the pathway marked in blue in Figure 3-4. When the NCA is released to federal agencies, how does the report affect their feelings, knowledge, skills, and actions? A necessary precondition is that the audience be aware of the NCA, leading to questions about awareness of the NCA within agencies. Then one can break down categories of outcomes (feelings, knowledge, and actions) into more specific areas of relevance (e.g., which specific parts of the NCA were relevant to the person's needs), which ultimately can be converted into questions appearing on a data collection instrument (e.g., interview protocol, survey questionnaire), producing measures that can be analyzed.

Identifying pathways important to USGCRP's goals also facilitates the integration of relevant theories and existing measures. For example, ideas related to the use of science in decision-making can inform the operationalization of concepts and the way they are measured. Whiteman (1985) theorized that there are two types of use of science in policy: (1) strategic use, which is using science to defend or support an already established policy, or (2) substantive use, which uses science to inform a policy decision. The connections between each portion of the logic model can be used to help determine what kinds of interrelationships are likely among items, setting up potential cross-tabulations, regression analyses, or other analytic tools. The next section discusses how the logic model can guide the formulation of overarching evaluation questions, which can then be elaborated further to investigate specific areas of interest.

Illustrative Evaluation Questions: What Will Be Learned from the Evaluation?

An important part of planning an evaluation entails evaluators engaging with the intended users of the evaluation to determine what they hope to learn from the evaluation. Working with the users will help produce a set of overarching evaluation questions that address the core of the information needed about the program (EvalCommunity, 2024). Notably, these are not the specific questions that would be included in a survey or data collection instrument; instead, they are the broad concepts that the evaluation will address. Six overarching evaluation questions are proposed below, as well as subquestions that further illustrate the concepts each overarching question seeks to address (Table 3-1). Even these subquestions would not necessarily be asked in data collection efforts; surveys or interviews will typically ask even more granular questions,[3] such as Can you provide examples of specific ways in which NCA products have helped you in your work? For each example, what specific parts were most helpful and why?

---

[2] Global Change Research Act of 1990, 15 U.S.C. Chapter 56, Public Law 101-606, 104 Stat. 3096-3104.

[3] To illustrate the difference between the questions used in data collection and the overarching evaluation questions/subquestions, consider Rodriguez-Franco and Haan (2015). That study used a survey with 64 multiple-choice questions as the primary data collection instrument. However, the overarching evaluation questions might be summarized as: What are Forest Service resource managers' perceptions of climate change? What are their science delivery needs related to climate change and its natural-resources impacts?

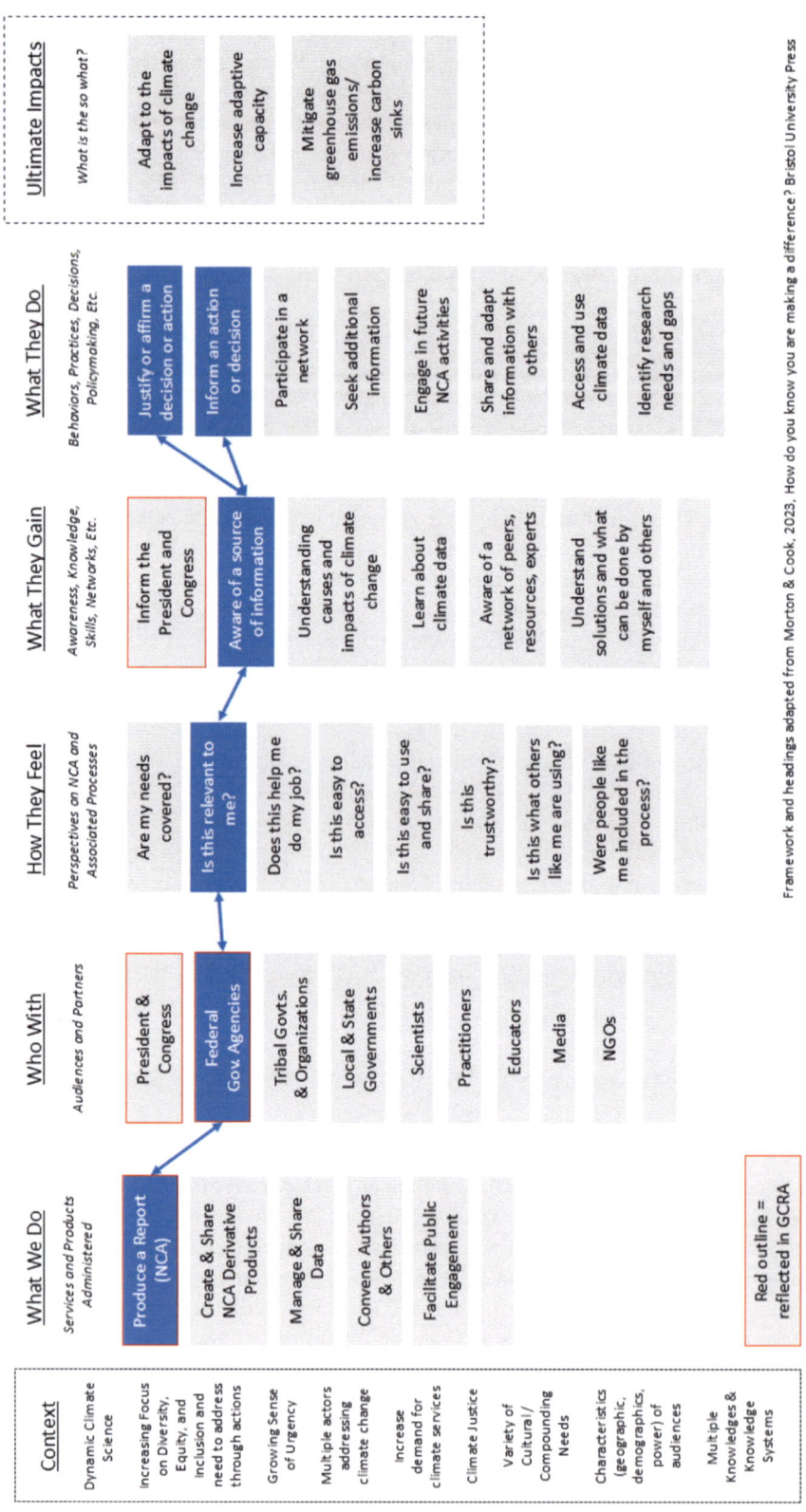

FIGURE 3-4 Sample user pathway through the logic model.
NOTES: A potential user pathway is mapped to explain how a person within a federal agency may use the National Climate Assessment (NCA). This individual would need to be aware of the NCA and its relevance before the pathway is followed in an action or decision. Pathways in a logic model help to define questions that could be asked and evidence to be gathered in an evaluation. For example, individuals in a given position or agency could be asked questions in a survey related to the relevance of NCA, its specific chapters, or derivative products. GCRA = Global Change Research Act of 1990; NGO = nongovernmental organization.
SOURCE: Generated by the committee.

TABLE 3-1 Preliminary Overarching Evaluation Questions for the National Climate Assessment (NCA)

Evaluation Questions and Subquestions

1. *To what extent are priority audiences aware of NCA products and what are the most effective ways to increase awareness? How, if at all, did involvement in the development process contribute to general awareness and use of the report?*
   a. For audiences that were not previously aware of the NCA, how did they become aware of the NCA? What factors, including the NCA development process, are related to awareness of the NCA?
   b. How, and to what extent, did users share information from NCA products with others? What could be done to make it easier for users to push information out to others?
   c. To what extent are audiences using other resources that are based on products of the U.S. Global Change Research Program (USGCRP)?
   d. What groups have been involved in the NCA development process and to what extent were they involved? How effectively were historically marginalized and underrepresented stakeholders included? What has been the impact of their involvement in terms of their later use and promotion of the NCA?

2. *How, and to what extent, did NCA products address information needs among priority audiences (i.e., what did they gain cognitively in terms of knowledge, skills, attitudes, capacities, etc.)?*
   a. What informational and other cognitive needs were well-addressed by the NCA?
   b. Which products or parts of products contributed to meeting information needs?
   c. What wasn't gained that could or should have been? What are the gaps in what the NCA provided and what else did potential audiences use to cover the gaps or for other purposes (including sources based on USGCRP products)? Are these gaps that audiences expect the NCA to fill?

3. *How, and to what extent, did NCA products address decision needs among priority audiences (i.e., what did they do as a result of using the products)?*
   a. What decision needs were well-addressed by the NCA? What decisions can audiences make given the level of information provided?
   b. Which products or parts of products contributed to meeting decision needs?
   c. What actions weren't taken because audiences did not have their decision needs adequately met? What are the gaps in what the NCA provided and what else did potential audiences use to cover the gaps or for other purposes (including sources based on USGCRP products)? Are these gaps that audiences expect the NCA to meet?

4. *How did the attributes of the products and process contribute to how users feel, what they gain (e.g., cognition), and what they do (e.g., behaviors)? What about the products and process could be changed to make them more effective?*
   a. How did engagement in the process affect use of the NCA? How did the engagement of historically marginalized communities and underrepresented audiences affect use of the NCA?
   b. How does the selection of topics, regions, sectors, level of detail, ways of expressing uncertainties, and other aspects of what information is included and the way it is organized and presented contribute to audiences' perceptions of how the products meet their needs?
   c. Given the abundance of other climate change resources, what is the unique value of the NCA and related products?

5. *How do the contextual factors described in the logic model influence how audiences feel, what they gain (e.g., cognition), what they do (e.g., behaviors), and how they mediate the use of the NCA?*
   a. How did the characteristics of the audience—including those who have been historically disenfranchised—affect the extent to which their informational and decision-making needs were met?
   b. How did the context of multiple information sources affect the impact of the NCA? Did they expand its impact? Did they replace it?
   c. Given the evolving nature of the field, what are future needs and how could USGCRP help meet them?

6. *How does the network of networks factor into use?*
   a. What organizations or types of organizations are the most powerful in spreading USGCRP products to other audiences (i.e., most influential nodes in the network)? What are the approaches that best support the sharing of information through the network of networks?
   b. How does USGCRP instruct or amplify the work that is being done by agencies and other partner organizations? How does the information generated by the NCA help facilitate their work?
   c. What is the part of this effort that USGCRP does directly? What parts are appropriate for the extended network to do?
   d. How did the NCA affect other information sources regarding climate change? In other words, do those information sources make use of it?
   e. Has the NCA affected the degree to which researchers and decision-makers collaborate? How does it affect the identification of research needs or development of additional research?
   f. Who is left out in the network of networks approach?
   g. How do frontline federal workers in these agencies feel, what do they gain, what do they as a result of the NCA?

Which NCA products or parts of products were not helpful? What was the result of your use of the NCA products? An example of a data collection instrument of this kind is in Rodriguez-Franco and Haan (2015).

Note that each of the six overarching questions has a different purpose. In order, these are awareness, information needs, decision needs, products and processes, contextual factors, and the network of networks. These topics have some overlap; for example, if a product is poorly organized (Question 4), that may affect the degree to which information needs and decision needs are met. Still, each question plays an important role. For example, if the NCA meets people's information needs but not their decision needs, then it is not fully meeting its purpose.

Working with evaluation users to develop overarching questions (and potentially priorities among them) is important because it places the focus on what those evaluation users hope to learn. This, in turn, is a key step in determining which methods are most appropriate to apply to which question. For example, if evaluation users are interested in learning the extent to which the NCA or other products are being used, then conducting broader surveys or leveraging existing data (e.g., citation analysis) may be more appropriate. However, if evaluation users are interested in gaining an in-depth understanding of how audiences use products, a series of interviews or focus groups, culminating in a small number of case studies, might be more appropriate. Because the overarching evaluation questions here are best answered collectively using a combination of methods, Chapter 6 offers guidance on selecting an effective blend of methodologies.

The questions in Table 3-1 are examples of the types of overarching evaluation questions that might guide an evaluation of the NCA. They are closely aligned with the logic model presented in Figure 3-3. For example, some of the questions below reference "contextual factors"—those refer specifically to the items highlighted in the far-left column of the logic model. Similarly, other questions refer to how users feel, what they gain, and what they do—those refer to Columns 4–6 of the logic model, respectively.

Because of the way the evaluation questions are structured, it is necessary to refer to the logic model when interpreting them. For example, Question 3 speaks of the effects of USGCRP products on the full range of actions that are described in "What They Do" (see Figure 3-3). As such, this question encompasses a wide array of potential actions, including making decisions about policy, additional research to conduct, and whether to participate in further USGCRP activities or networks. Because the Statement of Task did not provide insights about what types of decisions (e.g., decisions about how to reduce emissions or about adaptation or risk management strategies) or types of information (e.g., about tipping points or about the distribution of extreme risks) are of greatest interest, the evaluation questions do not include that level of specificity. The committee anticipates that USGCRP will often include greater specificity in the questionnaires or interview protocols. Some topics could be listed as individual items within a "check all that apply" type of question; others could be examined in greater depth. Because of this alignment between the logic model and the questions below, the overarching questions will need to be revised if the evaluators propose a logic model with a different structure. In addition, as discussed above, it is imperative that the evaluators work with the priority users of the NCA to determine what information the evaluation must yield to be useful. Therefore, the final overarching questions will differ from those proposed below because they will be informed through that discovery process.

The preliminary overarching evaluation questions are based on the logic model, the Statement of Task, and the discussions the committee had during its open meeting. These questions cover the same themes as those in the Statement of Task, but they have been reorganized and reframed, in part to reflect the logic model. Table C-1 (Appendix C) provides a crosswalk between the questions below and those in the Statement of Task, to demonstrate that all the concepts in the Statement of Task are incorporated.

The committee determined that to best understand how the NCA meets the needs of priority audiences, it is important to assess how information flows through the sprawling networks connected to USGCRP. A changing climate affects everyone, but the NCA may reach only a small fraction of people in the United States. Chapter 4 introduces network analysis, a branch of applied mathematics that studies nodes, such as organizations, and the connections that link them, such as the transmission of information. USGCRP is a central node in the interagency research network producing reliable climate science. Chapters 4 and 6 discuss how network concepts can be used to design an evaluation so that it illuminates how information in the NCA reaches those making decisions. And, as discussed in Chapter 2, USGCRP has at times employed a formal network of collaborating organizations, each of which is a node in its own network, to engage audiences and disseminate the information in the NCA. In these and other ways, USGCRP interacts with a network of networks in the transmission of climate science.

What Will Be Done with Evaluation Findings?

As emphasized throughout this chapter, a core principle of utilization-focused evaluation is ensuring that evaluation findings will prove useful. As such, it is critical to consider how individuals and organizations intend to use the evaluation from the beginning of the process, including in the development of the overarching evaluation questions and logic model and in the prioritization of evaluation participants (as described in Chapter 5). Chapter 2 suggests two modes through which the evaluation findings might be applied—to support continuous assessment and improvement, and to illuminate strategic questions raised by the broadening of the NCA's scope over time.

With regard to continuous improvement, there are many ways in which findings from the evaluation could be translated into recommendations for enhancing or refining USGCRP processes and products. For example, overarching evaluation Question 1 will yield information about how individuals become aware of NCA products and gaps in awareness. As USGCRP learns about the relative effectiveness of different dissemination mechanisms, it can shift resources from less-effective to more-effective tools. Similarly, by determining which parts of the network of networks are most effective in disseminating USGCRP products (Question 6), lessons can be learned from those entities that can be shared with others. This ability to apply findings from the evaluation on an ongoing basis will be important, given that the sixth NCA (NCA6) is already under development. Subquestions 3b and 4b probe for details about what aspects of the fifth NCA were most useful to audiences. Understanding those elements could directly affect planning for NCA6. In addition, as described in Chapter 6, the evaluation may help establish some practices of ongoing data monitoring (e.g., by setting up targets and systems to track media references) that could be employed by USGCRP on a continuous basis to inform changes in dissemination strategies.

A second mode of learning from the evaluation is focused on broader decisions about the direction of USGCRP and its products. Within this context, it is important to consider that there may be a distinction between the types of information that could be learned from an evaluation and how that information may be used. For example, as part of its work in encouraging researchers and policymakers to collaborate in producing the NCA, USGCRP has effectively broadened the scope of the NCA, making it not only a report for Congress and the president but also an arena for a broader community of researchers, decision-makers, and information users (see Chapter 2). The evaluation can provide *information* on how well that broadened scope has worked in practice (e.g., Question 4 will provide insights on what specifically makes USGCRP products effective to a range of audiences, while Question 6 will further clarify the mechanisms of usefulness). In addition, the evaluators could include questions in their data collection instruments to ask audiences their perspectives about USGCRP's scope and priorities (e.g., Questions 2c and 3c suggest that surveys or interviews might include questions about audiences' expectations of the NCA and USGCRP products).

Ultimately, however, the *use* of that information—in other words, determinations about whether that broadened scope is appropriate—will be in the hands of decision-makers at USGCRP and its sponsors. They may consider this information, along with values, resources, and constraints, when determining priorities and scope. When determining the overall evaluation questions, the evaluation team considers whether the knowledge that would be gleaned in answering them would provide sufficient information to inform decision-makers.

This discussion underscores the importance of establishing a logic model that comports with evaluation users' understanding of what USGCRP products are meant to achieve. If the ultimate impacts and the intermediary outcomes are not those with which decision-makers are most concerned, the evaluation will not provide usable information to meet evaluation users' goals. Similarly, the prioritization of USGCRP audiences and partners, which in turn affects who will participate in the evaluation (as described in Chapter 5), is key to ensuring that the evaluation will learn from those audiences who are most likely to influence the outcomes that matter.

Ultimately, an evaluation requires many applications of judgment; as such, it reflects an organization's culture and values. Clearly defining an ultimate impact and intermediate outcomes that demonstrate progress toward that impact involves making value judgments about what is important to an organization, program, or project—and what is not. The types of indicators selected, data gathered, and how those data are interpreted in an evaluation can also be colored by the values of the organization. For example, organizations with a propensity for quantitative data may choose an evaluation design that looks very different from an organization that favors more interpretive, qualitative types of data (Morton and Cook, 2023). Selecting which audiences will be a focus of the evaluation, and how they are engaged in the evaluation process, also reflects values. For example, the evaluation design could prioritize mutual benefit with communities currently underserved by the NCA, which could reflect USGCRP's values of addressing diversity, equity, and inclusion efforts. Reflexively considering values throughout the evaluation design can increase the likelihood that the evaluation leads to more relevant, meaningful, and actionable insights.

To increase the utility of the evaluation, the design process should begin by considering what information would best help evaluation users make decisions, on both the more process-oriented scale and the larger strategic one. Making sure the overarching evaluation questions, logic model, and the evaluation participants who are prioritized align with evaluation users' priority interests will help the evaluators establish the right methods to yield valuable insights.

CONCLUSIONS AND RECOMMENDATIONS

Conclusion 3-1: An outcome evaluation is an appropriate approach to addressing USGCRP's interest in understanding the extent to which its process and products have the intended effect on prioritized audiences.

Conclusion 3-2: Adapting well-established frameworks—such as utilization-focused, participatory, and equitable evaluation—from the beginning of the evaluation can provide standards and guidance to ensure that the evaluation will meet the needs of its intended users and incorporate equity considerations appropriately.

Conclusion 3-3: A logic model provides a useful conceptual framework for fostering a shared understanding of how activities are linked to outcomes.

Recommendation 3-1: The leadership of the U.S. Global Change Research Program should engage from the start in defining the evaluation scope and should ensure that the leadership perspective, as well as the necessary evaluation expertise, is incorporated throughout the design and implementation of the evaluation.

Recommendation 3-2: The U.S. Global Change Research Program should conduct an evaluation that can illuminate the outcomes of highest importance to the Program, including how the National Climate Assessment and related products inform significant decisions. The design and implementation of the evaluation should aim to be equitable and inclusive in both process and result, drawing upon the emerging knowledge of how to take diversity and justice into account.

Recommendation 3-3: The U.S. Global Change Research Program should develop a logic model to describe how its products, including the National Climate Assessment, are hypothesized to achieve their intended outcomes.

Recommendation 3-4: The U.S. Global Change Research Program should use a logic model developed for the evaluation to generate a set of overarching evaluation questions and should consult with partners and selected evaluation users to ascertain whether answering those evaluation questions will meet evaluation users' needs.

4

Network Analysis and a Network of Networks

Tracing and assessing the impact of the National Climate Assessment (NCA) and related products requires evaluation approaches that can not only make sense of these products' immediate usage, but also uncover how their influence ripples outward through indirect connections, as their contents spread throughout networks. The ideas of network analysis are helpful in conceptualizing that transmission of information. For example, by tracing the diffusion of information from the NCA, network analysis can help reveal the various audiences that ultimately utilized the report either directly or indirectly (and perhaps unknowingly through an intermediary) (see Appendix E). Network analysis can also help reveal information diffusion pathways that might support future dissemination efforts. For example, network analysis might help locate potentially underleveraged pathways or cultivate new ones (e.g., emerging opportunities using social media) as new platforms emerge and existing ones evolve over time (see Box 4-1).

Evaluators have used network analysis to understand the impact of interventions that reverberate throughout complex networks of affected organizations (e.g., Smit et al., 2020). By analyzing that network structure, an evaluator can help visualize the complexity of these connections and identify particularly important network connections or points of disconnection or weak connection (Popelier, 2018). This information can help to set priorities for the use of data collection methods, such as surveys and focus groups. In turn, an understanding of the network structure can help the U.S. Global Change Research Program (USGCRP) prioritize audiences and focus its outreach and engagement. In addition, an extensive literature applies network analysis to policy networks, including applications that focus on natural resource management and sustainability (e.g., Henry, 2020, 2023; Henry and Vollan, 2014; Henry et al., 2014, 2021; Weible and Jenkins-Smith, 2016; Weible et al., 2020), as well as studies on climate change science assessment networks specifically (e.g., Corbera et al., 2016; Venturini et al., 2023).

Although USGCRP has recognized the importance of networks for disseminating information from the NCA and its products, as evidenced by the creation of the NCANet more than a decade ago, it has not previously sought to explicitly incorporate the role of networks into evaluation of the NCA through network analysis. Because doing so would be a novel undertaking for USGCRP, the committee provided an overview of network analysis that may be useful as USGCRP considers incorporating this approach into any future evaluation. This chapter introduces ideas from the study of networks and suggests how they may support evaluation of the reach of the NCA and related USGCRP products.

Note that, as presented here, network analysis is a conceptual tool for examining the relationships between the many different groups engaged in climate-related activities, helping to illuminate how USGCRP's climate information is shared, known, and used. Network analysis is not a replacement for other evaluation methods;

> **BOX 4-1**
> **Information from the National Climate Assessment Spreads Along Networks**
>
> The Great Lakes Integrated Sciences and Assessments (GLISA)—a Climate Adaptation Partnerships program (formerly, the Regional Integrated Sciences and Assessments program) funded by the National Oceanic and Atmospheric Administration—provides an illustration of how information from the National Climate Assessment (NCA) is shared with wider audiences. The figure shows how that information moves from the NCA outward in a partnership with a nongovernmental organization, the Huron River Watershed Council (HRWC). Figure 4-1 depicts only outward transmission of information from the NCA; each of the nodes in the network also receives information from other sources.
>
> GLISA has deliberately enhanced its outreach by funding partnerships that might amplify its ability to connect climate information with users in targeted audiences (Lemos et al., 2014). In this example based on Kirchhoff and colleagues (2015), climate scientists at GLISA work with HRWC to engage with, translate, and tailor climate science for workgroups of officials—such as floodplain managers, water managers, and engineers—representing local governments. In doing so, a boundary organization (e.g., GLISA) takes in available information about climate change (e.g., NCA products), and connects it with constituents in communities throughout this watershed through connections with the HRWC and the local officials with whom the HRWC works. As the information from NCA diffuses through these layers, the immediate connection with the NCA products may be lost, but their influence continues indirectly.
>
> At every layer, relationships may affect the ultimate impact of NCA products, including relationships within and across networks among public officials and constituents. These chains might break, either because of a lack of connections or because a particular link is unreceptive to information about climate change from NCA products (because of climate denial narratives or other factors). The top route in the figure shows a flow of climate information denoted by the arrows from NCA products through HRWC and receptive local officials to constituents. In the bottom route, there is a break in the chain that cuts off this flow of information to whole networks of constituents when local officials are unreceptive. Network analysis may help reveal where such breaks might occur and where opportunities might exist to address them through targeted engagement or circumvent them using alternative networks.
>
>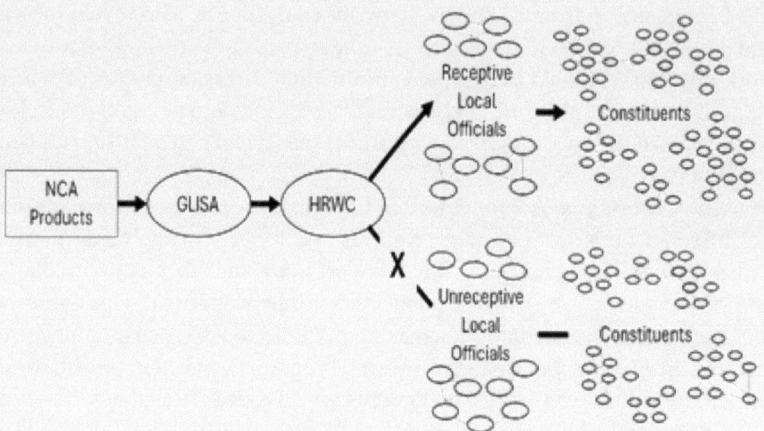
>
> **FIGURE 4-1** Transmission of information from the National Climate Assessment (NCA).
> NOTES: GLISA = Great Lakes Integrated Sciences and Assessments; HRWC = Huron River Watershed Council.

instead, it can play an important complementary role, identifying features of the way information is shared that can be investigated in greater depth by other means. Network analysis can inform and target the use of tools such as interviews, focus groups, and survey questionnaires to collect data about how information is accessed and used, about what works well, about difficulties encountered with respect to making use of the information, and about people's information needs and how or whether those needs are resolved. It can therefore provide insights that help inform the refinement of the kind of logic model laid out in the preceding chapter. At the same time, interviews, focus groups, and survey questionnaires can be essential for helping identify and measure influence and impacts rippling through indirect connections in these networks (Frank and Xu, 2020; Popelier, 2018).

NETWORK ANALYSIS

Network analysis encompasses a wide array of techniques for studying patterns of connections and how they give rise to outcomes of interest. A distinctive aspect of network analysis is its emphasis on the relationships between things in shaping what happens in the world. Although network analysis was already used prominently in technical work in industries such as communications and transportation, the emergence of the internet and social networking stimulated greater interest in social science applications of network analysis theory and techniques, with striking growth from 1995 to 2010, as online data and analysis software emerged (Brass, 2022). Many detailed and updated introductions are available (e.g., Borgatti, 2024, McLevey et al., 2023); this section has the more modest aim of providing background to inform discussions of how network analysis can support USGCRP product evaluation.

Network analysis features "nodes," which can represent all kinds of entities (e.g., individuals, organizations, documents), and "links" (or "edges"), which can represent different kinds of relationships between nodes (Box 4-2 provides a summary of some relevant types of connections). Network maps (or graphs) that visualize nodes and the links connecting them are a characteristic feature of network analysis. Originating in the mathematics field of graph theory, network analysis features methods that are generally applicable for understanding phenomena in almost any field, setting, and scale. Network analysis can be highly technical, but it can also be used in mixed-methods research, in conjunction with survey results or qualitative interviews that provide essential ground-truthing of findings or deeper understanding of the phenomena being explored. This flexibility means that network analysis can be used in evaluation in two ways: to gain a road view of how information is spreading and is used in networks, and to improve the efficiency of other methods, including surveys, interviews, and focus groups, so that they obtain more in-depth insight into how specific users interact with information and each other.

The NCA is cited in the communications of some of its important users. Citation networks afford quick understanding of the spread of some important kinds of information. In a citation network study, the nodes are documents and the links are citations made in one document that refer to another document. Academic research and other types of publication—including social network and website postings—are increasingly available in digital form on the internet. Network analysis tools and techniques have emerged to take advantage of these resources (see Kong et al., 2019, for a review). Assessing citation patterns throughout networks of documents and authors citing one another has become particularly prominent in the last couple of decades and can be used for applications

---

BOX 4-2
Example Types of Relevant Connections for Evaluation

A wide variety of connections within networks might be of interest for examination in an evaluation. Here are some examples:

1. Citations—In citation network analyses, papers that cite each other are linked. Appendix D provides more detail about the way citation networks can infer connections between sources of information and how information might flow and evolve between sources.
2. Coauthorship—Coauthorship networks connect papers or documents and individuals that helped author them. These networks can help with identifying how ideas spread over time as individuals participate with one another on different projects over time.
3. Weblinks—Networks of web-based sources can be studied via the weblinks connecting them. Similar to citation networks, weblinks can help identify how ideas spread between sources over time.
4. Media mentions—Media mentions of specific content can reveal connections beyond those who mention publications in their own communications.
5. Participation in meetings/events—Network analyses can be carried out when information is available on who attended meetings and events. Not only can such analyses help reveal ways that ideas might spread, but also events can provide opportunities to potentially intervene and build up new connections in networks.
6. Other social interactions—Much network data is collected by asking people who they interact or collaborate with. This can provide direct insight into specific kinds of relationships that are of interest and how these relationships might affect outcomes of interest, such as the spread and use of information.

such as exploring the structures of existing fields of study and understanding the impact of a particular piece of information, its diffusion, and patterns associated with its use (Zhao and Strotmann, 2015, p. 2).

Social network analysis techniques have also been widely utilized to understand social media networks (Bazzaz Abkenar et al., 2021; Saganik, 2019) and their impacts on outcomes through influence and information propagation (Chen et al., 2022). Many applications of network analysis rely on far smaller, more targeted data sources. Analyses of meeting attendees or selected documents can be used to understand how information or ideas might spread through networks and identify key nodes connecting otherwise disconnected parts of networks. For example, Braunschweiger (2022) analyzed Swiss adaptation policy development and implementation documents to define a network; they showed how the Swiss Federal Office for the Environment played a unique role as a national authority that connected work across sectors and scales. Interactions at meetings can also provide insight into diversity, equity, inclusion, and justice considerations, such as the marginalization or inclusion of diverse backgrounds and the extent to which efforts to enhance engagement of specific groups are successful (e.g., King et al., 2023; Vasquez et al., 2020). Engagement with key contacts who work with these communities could help identify events to include in a network analysis that could provide important insight into connections that influence the patterns of interactions of underserved audiences. Key contacts might also provide important perspectives about what sources of social media data may best reflect a targeted community's interactions, as different communities connect using different platforms. Appendix D provides additional details about these methods that may be useful in designing an evaluation of the NCA and related products.

SOME CONCEPTS AND USES OF NETWORK ANALYSIS

Network ideas can contribute valuable perspectives on the transmission of information, and these can be helpful in studying how climate science is transmitted.[1] Identifying the most important or influential nodes in networks is a common goal in network analysis. *Centrality measures* quantify and rank nodes based on their relative positions within the network. Analysts have developed a wide variety of these measures, to address different ways that the relationships between nodes might affect outcomes of interest; these are discussed further in Appendix D. An analysis of the centrality of federal sources in the transmission of attributions of severe weather events, for example, could provide useful insight about how decision-makers in many sectors perceive the legitimacy of climate information.

Cluster analysis techniques are commonly used to detect *clusters* (or "communities") within networks, which are subsets of nodes that are tightly connected with one another but are relatively isolated from other nodes outside of their group. A federal agency might be a cluster, sharing information that could also be of value to other clusters.

Identifying nodes that play a critical bridging role between clusters is associated with some of the most influential concepts in social network analysis related to information exchange. Granovetter (1973) argued that "weak ties" in networks between different groups were critical sources of novel ideas and opportunities within networks. Burt and Celotto (1992) emphasized the pivotal role of individuals who can bridge between "structural holes" in networks and bring together complementary information they might have.

Network analysis can be used to identify these holes and develop interventions to strengthen networks and improve management outcomes (Frank et al., 2023). Network analysis can also help identify and provide support for critical nodes that can bridge structural holes between historically marginalized groups and more well-connected parts of networks (Lopez Hernandez et al., 2022). Social media networks in particular can provide spaces where weak ties between those in historically marginalized communities can make supportive connections with one another that collectively help create and reinforce connections between them and others (Montgomery, 2018).

An example of these ideas at work is the demonstration by Vignola and colleagues (2013) that a regional extension office was a critical node, filling a structural hole for transferring information between larger-scale scientific organizations and more local organizations that were managing agricultural soil loss. USGCRP might

---

[1] This chapter provides a general discussion of network analysis and how it might be used. Some well-established texts on network analysis are Carrington et al. (2005), Newman (2018), and Wasserman and Faust (1994).

play a role in connecting clusters to one another—a way of advancing its mission that is different from its role as a central node.

The importance of such bridging nodes is related to the perception that these nodes are effectively holding disparate parts of the network together and that the network would fragment into unconnected components without them. Such a network is said to have low *redundancy*. Networks with high redundancy continue to remain connected even as nodes are removed. In a redundancy analysis, nodes are progressively removed to test how resilient the network structure is to these losses. Institutionalized climate services, such as the federal agency programs discussed in Chapter 6, may have higher redundancy than informal networks, such as social media networks, whose lower redundancy may be associated with a higher vulnerability to disinformation.

The ideas of Granovetter (1973) and Burt and Celotto (1992) have influenced studies of diverse social phenomena, including changes in employment, the formation of start-up firms, cooperation in social networks, social contagion, influence maximization, and the spread of social movements (Rajkumar et al., 2022). The relative importance of weak ties when it comes to information exchange and innovation is still an area of active inquiry and debate (Aral, 2016; Kim and Fernandez, 2023). It is clear, nonetheless, that network analysis can be useful for understanding diffusion of information and innovations as they propagate from node to node throughout networks.

Cunningham and colleagues (2016) assessed the diffusion of information about climate change from government officials to residents of Shoalhaven, Australia, based on questions about who they received information from and shared it with. The researchers were able to identify key nodes that spanned formal government networks and informal residential ones. Identifying such nodes can help in the design of interventions that attempt to alter or accelerate in desirable ways the outcomes that networks produce (Valente, 2012, 2017). Frank and colleagues (2023) used cluster analysis to identify a structural hole in a ravine management network and worked with a nonprofit group, the Alliance for Great Lakes, to develop a regional advisory group that was shown to help close this structural hole, enhance information exchange, and influence management's consideration about climate change.

There has been growing recognition over time that understanding how networks interact with one another is essential for understanding the behaviors of complex systems (Aleta and Moreno, 2019; Bianconi, 2023). Rapid growth in this approach to network analysis across many disparate fields and applications has resulted in a proliferation of approaches and terminology (Kivelä et al., 2014). *Multilayer network* is the inclusive term for these types of network systems, as they can be represented as a series of network "layers." The layering of multiple networks amplifies the dynamics being studied, because "cascades" travel from within and between networks.

In this literature, the term *network of networks* is used to describe one type of multilayer network. In a network of networks, the different layers feature different types of nodes, but there can be connections between these different types of nodes across these layers. As individuals, they may belong to different networks based on the sectors in which they work (e.g., federal, university, nongovernmental organization [NGO], state). Each person might be considered a node with a "work sector," and each sector might be considered a separate layer in the analysis. Through their collaboration, these individuals/nodes are also connected to other work sectors, or layers in the analysis. In this case, the network of networks analysis approach can help identify how key connections impact the spread of information both within a particular sector (within a particular layer) and between sectors (spread from layer to layer). Using a network of networks approach can play an instrumental role in organizing the use of other methods in an evaluation, as discussed in Chapters 5 and 6.

Multiplex networks are another type of multilayer network (Aleta and Moreno, 2019; Kivelä et al., 2014). In a *multiplex network*, the nodes are the same in each layer, but the different layers represent different iterations of relationships between them. Using the same example of climate change collaborators, these collaborators may interact with each other in multiple ways and at multiple times (e.g., at conventions, through conference calls, through listservs). Each interaction might be considered a separate layer. Viewing such a multilayer network as a multiplex network might be useful for diagnosing how different types of interaction (e.g., via committees, meetings) between nodes might have enabled or limited the spread and usage of information or how the network has evolved and affected outcomes over time. For example, such an approach might be used to gain insight into the effectiveness of NGOs in providing climate services to underrepresented populations.

APPLYING NETWORK ANALYSIS TO EVALUATION

To conduct an effective analysis of the reach of the NCA, the evaluator would need to focus the evaluation and identify the limits of the network investigation. That is, USGCRP's logic model would need to identify key audiences, taking into account their role in the networks where the NCA or a related product are being transmitted or used. Not every member of a network can be interviewed, nor can every chain in the link be investigated. Evaluators might determine the "highest-value" nodes that could help provide insight into how networks spread information and impact information use. Determining the highest value is in part a subjective judgment as to what will best meet the needs of USGCRP and the evaluation team. Highest value might be based on nodes that play key bridging roles or on other criteria described in Appendix D. The evaluation could then engage a selection of nodes of various types to obtain an understanding of the ways that different audiences and users are touching the NCA.

As described above and in Appendix D, citation analysis, weblink analysis, document analysis, or evaluation of the use of NCA and USGCRP products in social media can provide useful estimates of how information in these products spreads and has influence. These techniques might be used to illuminate what types of audiences use these sources, how they might use them differently, and what other sources of information are used concurrently to gain insight into how sources of information supplement one another.

A multilayer analysis in evaluation might be particularly useful for visualizing and assessing the indirect impact of NCA products (Robins et al., 2023). Approaching this as a network of networks multilayer analysis, where documents or authors are differentiated by sector, might be particularly helpful for visualizing and identifying ways in which information spreads between and throughout different sectors. Approaching it as a multiplex multilayer network analysis, in which each layer reflects activities in subsequent years, could potentially highlight the impact that the release of NCA products have, by making comparisons before and after their release. Alternatively, a redundancy analysis could also help quickly visualize the impact that products have, by demonstrating how connections splinter when NCA products are removed from the network.

Findings from a network analysis are likely to suggest to USGCRP and its evaluator the ways in which the logic model should be reconsidered. Viewing the potential influence of NCA products as a flow between and within different layers of networks, as illustrated in Box 4-1, emphasizes that this influence is shaped by the perceptions and decisions of individuals within these networks. Mapping networks might help define mediators of the NCA and other users that could support the definition of audiences in the logic model or even reveal unexpected pathways to pursue that weren't initially incorporated in the model. This can provide critical insights about audiences, partners, and connections between them for addressing the "Who With" aspect of the logic model. However, additional efforts might be needed to explore "How They Feel," "What They Gain," and "What They Do," which also influence the NCA's impact as it travels through nodes on these pathways.

Complementary approaches can shed light on users' perceptions and experiences that can help inform understanding of these aspects of the logic model, their connections between one another, and—ultimately—the impact of the NCA. For example, the *mental model construct* is an approach for exploring how conceptions that people hold about the world within which they are acting impact their responses to information, interactions with each other, and actions (Jones et al., 2011). Evaluators have used different qualitative and quantitative approaches to explore these conceptions and their impacts. Hoffman and colleagues (2014) used a network analysis to identify key connections between concepts in farmers' definitions of sustainable agriculture that shaped their participation in extension programs and adoption of sustainable practices. Research on policy networks has also emphasized that the shape of networks can provide insight into understanding "What They Do," as network locations can influence diffusion of information, learning, and actions (e.g., Henry, 2016). At the same time, while network structure can impact collaboration (e.g., Henry, 2023), participants' perceptions can also impact the choices they make about the relationships that produce the network structure (Feiock et al., 2012; Lee et al., 2012).

It is important to recognize that network methods present persistent challenges related to causality. Networks are inherently complex systems in which participants impact one another in overlapping and often indirect ways. Network analysts use approaches such as longitudinal data and sensitivity analysis to help interpret the impact of networks on outcomes. However, explanations about the effect of networks should also be judged based on how well they hold up relative to other causal explanations for observations, rather than on whether or not they offer definitive explanations of cause and effect (Frank and Xu, 2020). To that end, network analysis may also provide

> **BOX 4-3**
> **Illustrative Networks That Can Be Studied as Components of a**
> **Network of Networks Transmitting Knowledge from the NCA**
>
> A wide variety of groups (nodes) use information from the National Climate Assessment (NCA) and U.S. Global Change Research Program (USGCRP) to inform, interest, and/or activate the members of their networks. Groups that participated in NCANet and that provide climate services would be valuable for evaluation. Examples of the range of possible nodes include:
>
> 1. Nongovernmental organizations or nonprofits—for example, the Sierra Club, Rising Voices (Indigenous voices), Interfaith Power and Light (faith groups), the BlueGreen Alliance (labor unions), or the Union of Concerned Scientists
> 2. Professional organizations such as the American Society of Civil Engineers, American Public Health Association, or American Geophysical Union
> 3. Private-sector organizations such as the Cadmus Group or CASE Consultants International
> 4. Federal programs such as Sea Grant and the federally supported regional climate science-to-action networks—the U.S. Department of Agriculture's climate hubs, the National Oceanic and Atmospheric Administration's Climate Adaptation Partnerships/Regional Integrated Sciences and Assessments program, and the Department of the Interior's U.S. Geological Survey climate adaptation science centers (see Chapter 5)
>
> A full listing of the organizations that participated in the NCANet is provided in Appendix B.

insights for how to evaluate and discuss causality related to complex dynamics and outcomes that USGCRP is interested in, consistent with the discussion of contribution analysis in Chapter 3.

Network findings can also help to guide targeted efforts to explore subsets of the network of networks in greater depth. Box 4-3 provides a list of potential candidate audiences for investigation through narrower assessments. A detailed discussion of potential audiences and their prioritization is provided in Chapter 5.

General Approach

This section discusses a general approach to collecting and analyzing data on the role of networks in extending the reach of the NCA. As described earlier in this chapter, some of the appropriate techniques include citation analysis (examining who cites the NCA, and how that fits in a broader pattern of citations), study of weblinks (e.g., who provides links to the NCA, along with the pattern of other links that they provide, then expanding to a larger network of links), and social media (e.g., those who follow the NCA and reference it with hashtags). Similarly, although this would greatly expand the scope of the analysis, one might look at hashtags for prominent client science deniers, examining how people using such connections are connected to those referencing the NCA. With information on the structure of the network from automated tools, an evaluation can use interviews and survey questionnaires to learn how network organizers perceive their networks, whom they seek to reach, and what tools they use to reach their networks. However, when incorporating an analysis of the role of networks into an evaluation, careful consideration must be given to the resources required for this, since collection and analysis of network data can be time-intensive and therefore needs to be included in the assessment budget.

Evaluation Questions

Network analysis provides approaches that can help answer the evaluation questions proposed in Chapter 3. It can shed light on how information from the NCA and USGCRP products was shared. Evaluators could engage persons in significant nodes to provide insights as to whether they shared USGCRP information or products with their network and whether they used the NCA in planning or preparing products to share with their network.

Indirect Users

An important purpose of mapping a network's reach is to understand the indirect audiences that might be missed in other data collection or analysis activities. A network of networks analysis can provide an understanding of the indirect users and highlight gaps in audiences. In turn, this information will allow USGCRP to prioritize audiences—either those who are engaged or those who are missed—and develop specialized products and targeted outreach to serve them.

Underserved Audiences

The network perspective also provides a useful means of finding some of the audiences that are missed as climate information spreads. In mapping the network carrying the NCA and related products, an evaluation identifies actors in the nodes of the network. These persons understand the audiences in their network and can provide helpful insights on which of those audiences is not using climate science information in their decision-making, either because they do not find it useful or because they are not aware of sources such as the NCA and derivative products that use climate science information. People in these nodes are thus a significant audience because they understand where and how the content of the NCA is not getting through. In turn, the audiences that are identified by the key nodes can be studied so that USGCRP can gain an understanding of how information transmission dissipates as the NCA moves outward through networks.

Informants in nodes of the network are likely to have insight into their own reach and the gaps that they wish to fill. They could be asked questions including Who are their audiences? Which of their audiences could make use of information the NCA or derivative products in their decision-making? Why are they not doing so—are they unaware of the usefulness of climate science information? Do they believe that such information is useful? Informants in nodes can also be asked Whom do they wish they could better engage? What kind of participation do they have from vulnerable communities? Are there nodes within their network that engage directly with vulnerable and disadvantaged communities? Nodes' understanding of their networks can suggest changes in the NCA logic model by identifying gaps and stimulating further evaluation of efforts to reach groups that are not being reached but are important to USGCRP.

It is important to keep in mind, however, that the network approach does not identify all audiences that could use the information in the NCA in their decision-making. Audiences that are not known to the nodes within the NCA network will not be found by looking through that network. In practice, as noted above, an evaluation cannot reach all the nodes in the network used by USGCRP, which further narrows the audiences of nonusers that can be found. Given the challenges of finding those who are not connected, directly or indirectly, to a network, even the limited insights that can be gleaned by including important node actors in the evaluation may be helpful. Evaluators may follow up with nonusers who are identified initially to pursue further insight into others that they may be aware of who are currently not connected to the NCA's networks.

Prioritizing Audiences

As described in Chapter 5, USGCRP will need to apply a set of criteria to prioritize NCA audiences for inclusion in the evaluation. Network analysis may reveal valuable information about particular organizations' roles in transmitting knowledge from the NCA, and it is important to take this into account in setting priorities. If the goal is to understand all users of the NCA and related products, the evaluation can use a network analysis to assemble an audience that includes nodes of various types. For example, an evaluation might include one or more of each type of node discussed in Box 4-3. Representation would also be useful across nodes that are focused on climate or view it as secondary, that have large and small audiences, that serve different sectors or regions, that participate in the NCA process, or that use the NCA as a resource.

An illustrative example is a network analysis of the NCA's contributions to climate services. USGCRP has a direct role in coordinating federal climate services efforts (see Chapter 2) and is connected to the federal programs providing climate services in communities across the nation,[2] which are themselves nodes in the network

---

[2] Examples of climate-related federal programs include the U.S. Department of Agriculture's climate hubs, climate adaptation science centers, and the National Oceanic and Atmospheric Administration's Climate Action Partnerships/Regional Integrated Sciences and Assessments program.

of networks. Those federal programs offer an accessible path for evaluating the reach of USGCRP products, as the federal teams have working relationships that reach into the networks (see Chapter 6). Nodes outside of the federal system can also be evaluated. For example, the Union of Concerned Scientists uses NCA information (Declet-Barreto, 2024) to communicate with its members, collaborators, and vulnerable communities.

For each potential node evaluated, it will be necessary to determine its evaluability. Will the evaluation be conducted via surveys and interviews, or are there existing data that can be used? If so, what types of data exist (e.g., is attendance at events captured)? This is also an opportunity to establish plans for future data collection. For example, if a node does not have information on the number of users its work reaches, it can be provided with suggestions for collecting those data in the future. If they saw value in collecting such data, it might benefit future evaluations.

FINDINGS, CONCLUSIONS, AND RECOMMENDATIONS

Finding 4-1: Network analysis provides a variety of techniques for identifying key respondents from whom to seek additional information. Citation analysis and analysis of social media influence can be applied relatively quickly using digital data.

Finding 4-2: The study of networks of networks is an emerging area of research and interest in the field of network analysis; in applying network analysis, evaluators need to bear in mind that the use and interpretation of some results may be open to question.

Conclusion 4-1: Network analysis is a flexible set of techniques that support broad assessments of information flow and impact, as well as more in-depth investigations of user experiences.

Conclusion 4-2: Network analysis can be used to inform network interventions and can alter the outcomes they produce in desired ways.

Recommendation 4-1: In designing an evaluation of the National Climate Assessment or related products, the U.S. Global Change Research Program should make use of network analysis as a tool for addressing the evaluation questions related to understanding who key actors are, how information is transmitted across multiple entities, which entities serve as key nodes for disseminating information, and how the network of networks supports that flow of information.

5

Criteria for Choosing Audiences to Include in the Evaluation

As discussed in Chapters 3 and 4, there are a large number and variety of potential audiences of the National Climate Assessment (NCA). Many have highly specific and discrete needs that may not overlap with the needs of other audiences. These potential audiences may obtain information directly from the NCA, indirectly through an intermediary (perhaps without knowing that the NCA was the original source) and possibly with changes introduced by the intermediary, or information from sources other than the NCA. Some may not be examining information about climate change at all (e.g., they are too busy, uninformed, apathetic, or do not see the need). These factors greatly complicate the design of an evaluation.

There is no master list of all potential audiences, and, even if such a list existed, a random sample from that list would not be meaningful and generally would not contain enough people in each audience. Practically speaking, too, a data collection instrument that works well for one audience may not work well for another because of differences in needs and knowledge levels across the audiences. Thus, audiences will often need to be examined separately, and an evaluation will need to select a small number of audiences on which to focus attention. A staged approach, in which successive evaluations look at different audiences, can provide greater coverage than would be practical in a single evaluation.

This chapter provides general criteria for selecting audiences to include in an evaluation. The choice of audiences will be affected by the resources available for the evaluation, and there will be advantages and disadvantages for each audience that might be selected.

Selecting the audiences to assess in an evaluation of NCA use involves the analysis and consideration of the complexity and the multiple uses of the information provided by the NCA. The variety and range of applications of the information involve a diversity of users with different social, educational, geographical, and economic backgrounds, among other characteristics. The committee, in addressing this complexity, suggests that the designers of an NCA evaluation develop a logic model to visualize the NCA's relationships across different audiences, uses of information, and impacts, and that they use network analysis to establish the relationships among multiple audiences. This recognizes that the NCA operates in a network of networks context, which provides evaluators the opportunity to identify gaps, including marginalized populations (audiences) whose needs may not have been addressed. Prioritizing those populations in planning an NCA evaluation can provide the U.S. Global Change Research Program (USGCRP) with insights to ultimately help address gaps and find opportunities to promote social justice, equity, and inclusivity. Setting priorities for audiences to involve in an evaluation will help in engaging those users of the NCA who play an important role in USGCRP's logic model. The evaluation should include a focus on those whose decisions make a difference and those who play influential roles in the transmission of information from the NCA. This chapter articulates criteria for setting priorities, after discussing considerations that stem from the objectives of the evaluation.

AUDIENCES PROVIDING INFORMATION NEEDED FOR AN EVALUATION

In identifying criteria for selecting audiences, it is important to consider an evaluation as a whole, and what is needed is to gather perspectives from users and potential users. That means, in particular, that the selection of audiences accounts for the goal of supporting continuous improvement of the NCA. An evaluation is a project that is limited in resources and budget and reliant on the availability of audiences to participate or the information available about the audiences to be included. Criteria for prioritizing audiences for inclusion in an evaluation are discussed below, with particular attention to audiences that an evaluation will examine to understand the outcomes of the NCA.

Importance to USGCRP

The logic model should clarify for USGCRP which of its products' users play significant roles in the pathways that achieve its goals. Those users should be included within the evaluation with enough breadth and depth to understand those pathways. Two audiences deserve special mention. The Global Change Research Act[1] defines the NCA as a report to Congress and the president. It is accordingly logical for an evaluation design to include these two audiences among the priority populations to be engaged. As key audiences and users of the evaluation, it may also be useful to consult with them during the evaluation study design.

More generally, the logic model may suggest that a particular audience has a key role as a decision-maker or as a transmitter of information. To assess importance, USGCRP and the evaluator are encouraged to consider their hypotheses about which pathways will lead to the outcomes of greatest interest. Some users of the NCA make national policies, yet some policies of a limited scope may affect more choices on the ground, such as local urban planning rules to minimize flooding. Some outcomes may not focus on policy, such as raising awareness of climate impacts among underserved audiences and informing members of those groups of potential responses and resources for adaptation.

Some audiences are important because they act as intermediaries—nodes in the network of networks that transmit information to other nodes (see Chapter 4); examples include professional groups, educators, and news organizations. Chapter 6 provides examples of pathways where intermediaries play an important role.

Generalizability and Applicability

Any evaluation will be able to collect data from only a subset of potential audiences, and it is important to consider whether what is learned about each of these audiences and their use of the NCA can be applied more broadly. This is partly a matter of representativeness: are the audiences and uses investigated likely to be similar to other audiences and uses? Does the evaluation design collect information that provides insights that can be translated from those individuals who are included directly in the evaluation into an understanding relevant to at least part of the wider population? Generalizability is one consideration in determining priorities.

In setting priorities, it is useful to consider that the organizations already known to USGCRP may not be representative of the audience that the Program considers important to inform. Within a priority audience, the evaluator should seek to involve some respondents who have not already worked with the NCA, so as to minimize selection bias, as it is called in statistics.

As discussed in Chapter 6, however, the evaluation question being investigated affects how one considers generalizability. From a statistical perspective, *generalizability* refers to the applicability of the statistic to a population beyond the respondents. More generally, one might consider whether an evaluation finding has relevance to a broader population or broader class of events, even if a statistic cannot be considered to be representative. If the question is what percentage of an audience is aware of the NCA, and the method used is a question in a survey, one would want to know whether the survey estimate reflects the entire audience and perhaps the degree to which the statistic applies to other audiences.

---

[1] Global Change Research Act of 1990, 15 U.S.C. Chapter 56, Public Law 101-606, 104 Stat. 3096-3104.

If the question is how the NCA is used in a particular decision context, a case study may be informative, even if the narrative it produces is not readily applicable in other situations. In particular, case studies of unusual success may offer insights relevant to the logic model, as well as suggesting lessons learned that might assist in replicating their success. Similarly, a finding that the online NCA report is difficult for some to navigate for a particular reason can be useful whether or not one is able to estimate the number who would benefit from a change.

When focusing on audiences, generalizability is defined in terms of the ability to generalize to a particular audience, not to the overall population. For example, suppose that an evaluation included a case study of a particular nongovernmental organization (NGO), such as the Union of Concerned Scientists (discussed in Chapter 6). At the most basic level, one would want to know the degree to which the case study covered the entire organization, or only a few individuals within it. It is not necessary for the results to be generalizable to the full group in order to be useful, but this is one factor to consider. Next, one might ask whether at least some of the findings would apply to other NGOs as well. Few results can be generalized to describe the entire universe of NGOs, because NGOs collectively are so diverse in their purposes, structures, and operations. There would be no realistic way of generalizing still further to the entire United States because the nation is much more than and much different from an aggregation of NGOs. Nonetheless, one may still find principles that are broadly applicable, such as that some audiences might need a much higher level of specificity—such as localization—than is contained in the NCA. As a corollary, the questions to be addressed sometimes will be audience specific, and an evaluation will not apply the same questions to every audience; for example, some audiences may be users only and not disseminators of NCA information.

A related practical consideration relates to sampling (Lance and Hattori, 2016; Palinkas et al., 2015). To understand the extent to which a phenomenon exists—for example, what proportion of local health department officials are familiar with the NCA—the evaluators would seek to survey all or a representative sample of those health officials. The feasibility of conducting a survey that will produce generalizable results about all U.S. local health departments depends on (1) how feasible it is to gather a list of all local health departments (the universe of local health departments); and (2) the ability to field a survey with adequate response rates that reflect that full universe. (As described in Chapter 6, one strategy for addressing these concerns is adding questions to existing surveys that already have well-established sampling processes and robust response rates.)

Diversity

One aspect of evaluation coverage deserves special attention: the inclusion of marginalized or underserved audiences. The needs of some populations have not been considered consistently in the development or provision of climate services, and they may lack access to the NCA or to decision-makers. Engaging their perspectives and understanding their needs could make the NCA a much stronger product. In gauging the size of this population, it is important to consider the definition of *underserved*. Many may not be aware of the NCA but still use its information or are affected by it. To the extent that climate information traceable to the NCA is reflected in public policies—such as development guidelines for floodplains or in planning for cooling centers during heatwaves—the NCA may be said to touch some of the decision-making and experiences of a wide range of residents of the United States. The specification of what counts as use of the NCA will affect the population estimated to be underserved. Such a definition is one that USGCRP could articulate as it develops its logic model.

Drawing upon research in environmental justice that identifies communities and demographic groups that lack access to climate services or climate information (Dolšak and Prakash, 2022; Tripati et al., 2024), the evaluator could propose an evaluation instrument to illuminate how climate information reaches and is used by underserved groups. As discussed in Chapter 4, a mapping of the network along which the NCA is shared can identify key nodes; the persons at those nodes are likely to have information about the audiences they are trying to reach and the limitations they encounter in doing so. As noted earlier, information gathered in this way relates to only a portion of those who do not use the NCA.

In addition, some people, including climate skeptics, do not believe that federal climate science is useful. Better understanding this population could provide information useful in improving the NCA and climate services in the future. In a recent study, the Pew Research Center interviewed climate skeptics and observed the way that mistrust of the news media interacted with reluctance to believe in climate change (Pasquini et al., 2023). Those who are not well served by existing climate services programs may also be affected by their access to, and trust in, news media; these and other interaction effects could be studied using small groups.

The Pew Research Center's studies indicate that climate change is perceived to be an important issue among U.S. residents, but one not as pressing as other problems, such as the state of the economy or health care (Tyson et al., 2023). Moreover, opinions about climate change are polarized, with Democrats readier to believe that climate change is important enough to warrant action by government, while Republicans are more cautious in their support. Substantial majorities do support climate mitigation policies, however, such as the development of renewable energy (Tyson et al., 2023).

What is more germane to the NCA is the modest confidence of the public in climate scientists. Pasquini and Kennedy (2023) introduce their article by saying:

> Only about one-third of Americans think climate scientists understand very well whether climate change is happening. . . . And only about a quarter or less say climate scientists understand very well the effect climate change has on extreme weather, its causes and the best ways to address it.

These findings suggest that many would not find the NCA useful even if they knew of its existence.

An evaluation exploring these beliefs would likely clarify what is expected of the NCA, as well as prompt ideas about how climate information could be more useful to selected audiences. The committee recognizes, however, that there are serious methodological issues in trying to conduct such a study, both in identifying the audiences to be studied and in judging how representative the results would be. What may be feasible is to conduct case studies or focus groups with a limited number of people in such audiences to better understand what information they have and what difficulties they face.

There is another population of interest: those who are aware of the work of USGCRP but do not find the NCA and associated products useful. Such people may be difficult to identify; it is easier to identify users (e.g., through monitoring downloads or citations) than nonusers. For these reasons, it may not be practical to specifically target nonusers. Instead, one might target groups based on their potential for benefiting from the NCA (e.g., target urban planners in general rather than urban planners who make use of the NCA), and then ask about their knowledge of and interest in climate science information, and their sources of information. Since knowledge and use of the NCA and its products might depend on the prevalence of climate services (provided by the federal government or other groups), useful information might also be gained by comparing individuals within a given community of practice (such as urban planners or public health officials) who do have ready access to climate services with those in that community who do not. Such a comparison might help in identifying not only how climate services have promoted use of NCA-derived information but also groups for which outreach might be particularly useful.

FEASIBILITY OF EXAMINING PARTICULAR AUDIENCES

In setting priorities for audiences and methods to apply in an evaluation, practical considerations such as identifiability, data availability, accessibility, and cost affect the options for evaluation design, and for how an evaluation is conducted or phased.

Identifiability. Although Congress, the president, and federal agencies have known and clearly identifiable memberships, other important audiences do not. Consider that, among others, there are many NGOs, only some of which are involved in climate change; there are many educators, but there is no national list to sample from and no way to screen them by their relationships to climate science information; and there are diverse media, ranging from traditional outlets to a wide variety of blogs and other sources on the internet. In those cases, generalizability of evaluation findings does not mean representativeness in a statistical sense. It will not always be feasible or practical to develop a comprehensive list or to generate statistics that can be considered representative; instead, one may want to focus on a particular subaudience (such as a specific NGO) and seek findings that may be applicable more broadly.

Data availability. Some data may be readily accessible to an audience that is being studied, such as the amount budgeted for climate-related activities or the number of individuals within an organization engaged in climate-related activities. In other cases, data may be available only with significant effort or not at all. These considerations affect which audiences to engage, what information might reasonably be obtained from them, and how reliable the information provided may be. An evaluator compiling secondary data (existing information that was not collected for the purpose of the evaluation) will need to assess the quality of those data and their relevance to the overarching evaluation questions.

Accessibility. Members of some audiences are less accessible than others. Interviews may be difficult to arrange, permission to reach out to subjects may not be obtainable, and securing the return of surveys or other measurement instruments may be uneven or impossible.

Cost. Budgetary and other resource limitations may make desirable audiences difficult to engage, forcing trade-offs between audiences or reductions in sample sizes.

EVALUATION STAGING

As is implied by the history of the NCA (Chapter 2), the broad relevance of the climate science in the NCA for decision-making (Chapter 3), and the complexity of the network of networks (Chapter 4 and considerations above), setting priorities to select audiences to include in the evaluation needs to engage with USGCRP's goals. It is a practical necessity to proceed in stages.

The first stage is an evaluability assessment, discussed in Chapter 7. That includes creating an initial roadmap of the networks believed to be most important to achieving the Program's purposes. In mapping the NCA's network of networks, using low-cost methods such as citation analysis complemented by case studies, the objective is to clarify the logic model and the evaluation questions to be answered. In particular, network analysis is likely to identify structural holes: portions of the NCA network that are connected to one another by only a few intermediary connections. These intermediary nodes, and the way that they provide connections, may play a significant role in the spread of information from the NCA. Some of these may be worth highlighting in the logic model with evaluation questions developed to examine the role of these boundary-spanning nodes. Chapter 7 discusses this initial reconnaissance in greater detail.

Insights from an initial mapping can then be combined with the considerations discussed in this chapter, in order to choose a set of audiences to include in an evaluation. Those audiences will need to have the appropriate breadth and depth to answer the evaluation questions, while also being practical to work with. In developing a roadmap, an evaluator needs to be aware of the way the methods used in the reconnaissance provide only a limited view of the network of networks; as discussed above, exploring why potential audiences do not use the NCA requires methods that engage audiences unconnected to the network of networks. The initial roadmap is only a starting point that provides information to pose questions to audiences in a way that can provide reliable answers to important evaluation questions.

CONCLUSIONS AND RECOMMENDATIONS

The discussion above can be summarized in the following conclusions and recommendations, which set the stage for the discussion of methods in Chapter 6. Each conclusion pertains to a criterion for prioritizing audiences:

Conclusion 5-1 (Importance to USGCRP): The legislative mandate, logic model, and evaluation questions can be used to designate key audiences.

Conclusion 5-2 (Decision-making): (a) Where the logic model focuses on outcomes of national scope, audiences that influence or make decisions at national scale are the most relevant. (b) Where the logic model focuses on outcomes that matter because they result from many decisions made in a decentralized fashion, audiences are likely to be numerous and diverse.

Conclusion 5-3 (Information transmission): Some audiences may be prioritized for inclusion in an evaluation because they are hypothesized to be key nodes in transmitting climate information. Existing communications networks such as the mass media, as well as communications within professional organizations or communities may deserve priority where they play influential roles in the logic model.

Conclusion 5-4 (Generalizability): When feasible, there is value in choosing audiences for whom the results would be generalizable (i.e., the evaluation findings for an audience or subgroup within an audience are likely to be applicable to other audiences that are similar in a relevant dimension). Generalizability is not always feasible depending on the nature of the audience.

Conclusion 5-5 (Matching audiences to questions): In many instances, audiences need to be matched to the question to be investigated and comprehensive coverage or representative sampling may not be relevant or feasible.

Conclusion 5-6 (Diversity): Targeting audiences that are collectively diverse is important for two reasons: (1) to provide a more complete picture of how the NCA is or is not used, and (2) as a matter of equity, some marginalized audiences may lack access to the NCA or to decision-makers, but consideration of their needs can make the NCA a stronger product.

Conclusion 5-7 (Feasibility): Because of cost and other practical limitations, not all audiences can be evaluated using the most powerful evaluation methods. Sometimes these audiences can be examined in a more limited but still useful way using different methodologies.

Recommendation 5-1: In choosing which groups to study as part of an evaluation, the U.S. Global Change Research Program should seek diversity (including a focus on marginalized populations) similar to that of the participants and audiences with which the Program seeks to engage.

Recommendation 5-2: The U.S. Global Change Research Program (USGCRP) should select audiences to include in evaluation based on the following criteria: importance in USGCRP's logic model, including (1) the role of an audience in climate-related decision-making; (2) the role of an audience in the transmission of climate information for decision-making; (3) the generalizability of results from an audience to other populations; and (4) feasibility, diversity, and suitability for the evaluation question and method used. A targeted audience does not need to meet all of these criteria, but the audiences prioritized in an evaluation should meet these criteria collectively.

Recommendation 5-3: The U.S. Global Change Research Program (USGCRP) should continue to pay attention to how climate change affects historically marginalized communities and underrepresented audiences by continuing to provide pathways for including them as part of the development process of the NCA, making sure that they are heard throughout the development process—and more broadly by sustaining efforts to provide information about the climate-associated needs of those groups. The evaluators engaged by the Program should include persons whose backgrounds and lived experiences afford them understanding of underserved communities, including those exposed to the impacts of climate change.

Recommendation 5-4: Guided by its logic model, the U.S. Global Change Research Program (USGCRP) should progressively develop a roadmap of the network of networks in which the NCA and its related products are used. This roadmap would show the nodes that are hypothesized to play key roles in the diffusion of usable knowledge in the NCA and related products. The results should be considered when selecting which audiences to target in the evaluation.

6

Methodologies and Applications to Particular Audiences

Chapter 5 discusses how different audiences offer different benefits to an evaluation and different levels of feasibility for investigation. It offers criteria for prioritizing audiences and considers how the differences among audiences require different evaluation methodologies for specific audiences. This chapter discusses evaluation methods, with a particular focus on methods for examining specific audiences, building on the discussions in Chapter 4 on the ability of network analysis to identify links connecting audiences and the properties of those links as a connected system.

To develop a methodological framework for an evaluation, it is critical to align the following: (1) what the evaluation seeks to learn (as articulated in the logic model and overarching evaluation questions, using the illustrative examples from Chapter 3), (2) from or about whom you need to hear to answer those questions (based on prioritization using the criteria from Chapter 5), and (3) the attributes of different evaluation methods. This chapter begins by providing a very brief overview of some of those methods and offering considerations for how different methods may be applied in combination to answer overarching evaluation questions. This is followed by considerations for tailoring those methodological approaches to priority audiences. Given the large number and diversity of audiences, no single approach will be consistently effective and feasible. This chapter discusses why customized strategies are needed and gives illustrative examples of how to develop customized strategies.

OVERVIEW OF POTENTIAL METHODOLOGIES

This section distinguishes between collecting primary data (e.g., surveys, interviews, and focus groups conducted specifically for the evaluation) and secondary data (e.g., data collected for other purposes that may be used to help answer overarching evaluation questions). After providing a brief overview of selected methods and their advantages and disadvantages, the chapter discusses considerations specific to evaluations conducted by the U.S. Global Change Research Program (USGCRP). *Case studies*, which may combine primary and secondary data to explore a particular instance in depth, are then highlighted as particularly relevant to USGCRP evaluations.

While this section describes each method separately, it is important to note that they are often deployed in combination. For example, network analysis (as described in Chapter 4) might identify subjects for case studies. Evaluators may conduct a survey first and then follow up with more in-depth interviews to gather nuanced information about survey findings—or they might conduct interviews first to refine what questions to ask in a survey.

(More information about sequencing of different evaluation methods can be found in Chapters 7 and 8.) Using multiple methods in combination is also consistent with the iterative approach of building support for contribution analysis, as it provides multiple opportunities to compile evidence to build, test, and revise the understanding of the link between cause and effect (Mayne, 2011).

This section concludes by drawing connections between the overarching evaluation questions from Chapter 3 and the methods described here.

Collecting Primary Data

Chapter 5 articulates a series of criteria that can be used to prioritize which participants should be incorporated into an evaluation to gain insights about their use of the National Climate Assessment (NCA) and other USGCRP products.

Overview of Collecting Primary Data

Mechanisms for collecting information directly from those prioritized audiences (primary data) include surveys, interviews, focus groups, and observations. Paradis and colleagues (2016) describe the "ideal" use of each of those methods:

- Surveys: to capture perceptions, attitudes, beliefs, or knowledge of a defined sample of individuals. Surveys are often also used as a data collection tool for large populations.
- In-depth interviews: to capture accounts of, perceptions of, or stories about situations or phenomena from individuals.
- Focus groups: to understand phenomena when "the sum of a group of people's experiences may offer more than a single individual's experiences" as well as to document how participants react to what they hear from others.
- Observations: to document what individuals actually do, rather than to ask them about their perceptions or recollections about what they do.

In addition to identifying which of those purposes best addresses the evaluation questions, there are practical considerations with respect to the different modes of data collection. Table 6-1, which draws from the Centers for Disease Control and Prevention (CDC, 2011) and Taherdoost (2021), highlights some of the advantages and disadvantages of each of these four methods. The table is not meant to be exhaustive, nor is the list of data collection methods. Instead, this table is meant to illustrate the types of trade-offs that will be encountered in determining an appropriate combination of methods. Evaluators can provide more in-depth guidance about the benefits and limitations of each method and suggest other possible data collection methods.

Considerations for USGCRP Evaluations

When determining which data collection methods to apply, it is important to begin by grounding the decisions in what evaluators are seeking to learn, as captured in the overarching evaluation questions and their associated subquestions (Chapter 3). This will help evaluators identify which method(s) are most likely to yield answers to the evaluation questions.

For example, surveys may be best suited to contribute to understanding the level of awareness about the NCA from a large number of audience (or potential audience) members, which relates to Evaluation Question 1 ("To what extent are priority audiences aware of NCA products and what are the most effective ways to increase awareness?"). On the other hand, to understand *why* priority audiences take action, interviews may provide the opportunity to hear the story behind a decision to act, which relates to Evaluation Question 4 ("How did the attributes of the products and process contribute to how users feel, what they gain (e.g., cognition), and what they do (e.g., behaviors)? What about the products and process could be changed to make them more effective?").

TABLE 6-1 Selected Advantages and Disadvantages of Data Collection Methods

Selected Advantages	Selected Disadvantages
Method: Surveys	
• Feasible for reaching larger groups of individuals in a time- and cost-efficient manner, which presents greater potential for generating information that is generalizable across other members of the group sampled in the survey • Can cover a large range of topics • Possible to collect anonymous data (doing so might increase likelihood of candid responses)	• To the extent that surveys use forced-choice questions, they may miss potential responses and nuances. On the other hand, standardizing responses helps when tabulating responses • Special care is needed to ensure wording does not bias responses
Method: Interviews (individual/in-depth)	
• Can gather greater breadth and/or depth of information • Offer opportunity to ask for clarification or follow-up based on responses • Provide opportunities to identify concepts or pretest questions for further use in surveys or other data collection exercises	• Conducting in-depth interviews is more time-consuming and expensive than conducting surveys • Resource constraints could limit the number of individuals interviewed, making it less likely to yield information that would be generalizable across a wider group • Process of coding and analyzing data is time-consuming • Bias remains possible depending on what questions are asked and how they are worded.
Method: Focus Groups	
• Can help discover social concepts, develop hypotheses, and uncover commonalities and divergent viewpoints • Allow participants to react to comments from others	• Require skilled and trained facilitator(s) to control the extent to which all participants can contribute to the session • Can be logistically challenging to coordinate multiple participants • Conducting focus groups is more time-consuming and expensive than conducting self-administered surveys • Not designed to produce quantitative data or data that will be statistically representative
Method: Observations	
• Can provide more detailed information about how a program is implemented	• May be time-consuming • May be difficult to interpret observations • The process of observing may change how the activity is conducted or how people respond

Given their interactive nature, focus groups may be particularly valuable for informing how the NCA is viewed by audiences within the context of the broader climate information ecosystems, per Evaluation Question 5 ("How do the contextual factors described in the logic model influence how audiences feel, what they gain [e.g., cognition], what they do [e.g., behaviors], and how they mediate the use of the NCA?"). Having a group of participants discuss their varied sources of information about climate in a focus group might produce a more comprehensive list of alternative sources and therefore foster greater discussion about how NCA compares with and complements those other sources.

Evaluation Questions 4 and 5 can provide insight into how historically marginalized communities engage with the NCA and its development process. During its open meetings, the committee heard that simply inviting individuals to join part of the NCA development process may be insufficient for truly engaging them in the discussion. Observations might yield valuable insights about how diverse participants engage in NCA deliberations and development and what opportunities might exist to foster truly inclusive discussions. Based on the observations, selected individuals might be interviewed to understand more about whether their viewpoints were properly represented, and what barriers or encouragements they may have encountered.

While evaluations that seek to answer multiple complex overarching evaluation questions will likely seek a mix of different methods, that specific mix will depend both on the questions to be answered and on practical considerations, including some of the concerns raised above.

Considering the feasibility of fielding surveys with adequate and representative responses, it may be appropriate for the evaluators to seek to add questions to existing surveys of the prioritized group, rather than fielding a separate survey. This will also help to reduce survey fatigue by limiting the number of surveys being sent to the same group of individuals. This method depends on finding a survey that is targeted to the desired population and on the ability to ask questions of sufficient depth in a survey that may be designed for a different purpose.

For in-depth interviews and focus groups, which are seeking to answer questions more about how and why, rather than how many, the need to have a statistically representative sample may be less significant than gaining insights on the mechanisms behind understanding, decisions, or actions. Rather than using a statistical approach to sampling, qualitative data collection often uses "purposeful sampling," which seeks to identify "information-rich cases for the most effective use of limited resources" (Palinkas et al., 2015, p. 534). This concept is further explored in the discussion of case studies below.

Document Review and Analysis of Secondary Data

In addition to collecting new data, compiling and analyzing existing data can yield important insights in evaluation (Vartanian, 2010).

Overview of Using Secondary Data

Secondary data are those that were collected for a purpose other than the evaluation at hand. They can include program reports, data routinely collected by organizations for other purposes (e.g., attendance at webinars), media analysis, and information in the scientific literature, among others. Some of the advantages of using secondary data include avoiding duplication and the potential for saving time and money. Secondary data can also provide additional context for primary data and therefore a broader understanding of the question at hand. As such, secondary data analysis is often used to complement primary data collection to address overarching evaluation questions more holistically. However, because secondary data are collected for other purposes, the validity of the analysis will depend on the quality of those data sources. What is more, the concepts and measurements used in secondary data may not exactly match those that are the focus of the evaluation (McCaston, 2005; NCVO, n.d.). Secondary data may already exist in a readily usable form (as in a general survey that included questions about climate change) or may require additional compilation or processing to prepare the data for analysis.

Considerations for USGCRP Evaluations

The logic model can help inform hypotheses about the types of documents that are most likely to demonstrate NCA's contribution to decision-making and help determine what types of databases might be most valuable to search. For example, an evaluation team may theorize that use of the NCA or related products in the development of policy documents is a critical pathway for informing decision-making. If so, evaluators would specifically tailor their search approach to capture policy documents.

Textual review of existing documents can be used to complement primary data collection. For example, an evaluator may review the websites of a handful of U.S. conservation and environmental organizations to determine which ones appear to be using USGCRP products, to identify organizations for initial interviews. Depending on the final set of overarching evaluation questions, the evaluators may seek to reach out to a selection of both those organizations that appear to use USGCRP products (to learn more about their use) and those who appear not to use it (to learn more about underserved populations and their reasons for not using those products). Conversely, if an evaluation were to begin by administering a survey to public health officials, the evaluation team might use those responses to identify examples of health departments that are actively engaged in climate change work and use that information to focus their review of websites or documentation on those organizations that have indicated engagement in this area.

Some forms of secondary data analysis may be particularly beneficial to support continuous improvement. Web analytics of USGCRP pages could not only help identify which parts of the NCA and related products are most frequently being viewed, but also provide insights on user experience with the website (Palomino et al., 2021). Similarly, web scraping might be used to gather information about where the NCA is cited or mentioned on the web in general or within particular communities of practice (Mitchell, 2018). This might inform decisions on areas of focus and web and interaction designs for future NCA reports. Similarly, there is a growing body of tools and frameworks for tracking how social media and news media are spreading content (Almasy and Thompson, 2013; AMEC, 2024). An evaluator might help USGCRP establish targets and systems for ongoing tracking in these areas, to support continuous improvement of USGCRP outreach efforts. Given the emphasis on diversity, equity, and inclusion, it would be helpful to collect demographic information, where possible, from individuals who register for USGCRP events or sign up for the newsletter.

Case Studies

The U.S. Government Accountability Office (GAO, 1987) defines a *case study* as "a method for learning about a complex instance, based on a comprehensive understanding of that instance obtained by extensive description and analysis of that instance taken as a whole and in its context" (p. 9). Put another way, case studies provide in-depth information about a particular organization or scenario, often drawing from collection and analysis of data from a variety of sources (Balbach, 1999). For example, evaluators may analyze program documents, conduct a survey, and engage in multiple interviews all about one organization. This methodology may be especially relevant for USGCRP, as it allows an evaluator to gain an in-depth understanding of the mechanisms by which something occurs. As Yin (2009) explains, "[t]he more that your questions seek to explain some present circumstance (e.g., 'how' or 'why' some social phenomenon works), the more that the case study method will be relevant" (p. 4).

While some researchers emphasize that a focus on a small number of cases limits the ability to create findings that can be used to make statistically significant statements about the impact of a program (one understanding of the word *generalizability*), the case study method is particularly useful for illuminating the theory behind the intervention, and that can help build understanding of how or why a program may be effective (Tsang, 2014; USAID, 2013). As such, case studies may be especially effective in drawing lessons learned and information that can be applied to improving the intervention—or, in this case, supporting continuous improvement of USGCRP products and processes.

Multiple types of case studies (USAID, 2013) might be applicable to USGCRP. A *critical instance case study* describes why a specific case is unusual. In network analysis, this type of analysis may be helpful to explore why and how a particular node appears to be so influential. For example, it might be appropriate to conduct a case study of a nonprofit organization that appears to be especially effective in translating information from USGCRP into formats used by its members or constituents. Determining what makes that organization successful could yield information relevant to other organizations. This aligns with a positive deviance approach, which uses statistical analysis to identify cases that perform better than their peers with similar characteristics. Focusing case studies on those outliers can reveal insights about what makes some organizations particularly effective (Klaiman et al., 2016).

An *explanatory case study* investigates the effects of a particular program and seeks to make causal inferences about the success (or lack thereof) of the program. For example, USGCRP could conduct such a case study of a climate-relevant policy that was enacted by a state or local governmental entity to understand how USGCRP contributed to that outcome and what worked (or did not work) well. Such findings can be applied to continuously improve USGCRP processes and products.

Comparative case study designs look at multiple cases to identify similarities and differences. This type of approach could be used, for instance, to see how different federal climate programs have used USGCRP products and could yield important insights on the network of networks approach.

Connecting Methods to USGCRP Evaluations

The range of methods selected in an evaluation always needs to be based on what the evaluation is seeking to learn. As such, Table 6-2 provides methodological considerations for each of the illustrative overarching evaluation questions listed in Chapter 3.

The left column provides examples of how those illustrative overarching questions, which broadly describe what can be learned from the evaluation comprehensively, can be translated into questions that are specific enough to be included in a survey or interview/focus group guide. This list is not exhaustive, but it illustrates how the overarching questions, as well as specific elements from the logic model, are translated into more granular questions.

TABLE 6-2 Data Collection Methods and Considerations for Overarching Evaluation Questions

Primary Data Collection: Examples of Questions for Surveys, Interviews, Focus Groups	*Considerations for Analysis of Both Primary and Secondary Data*
Evaluation Question 1: To what extent are priority audiences aware of NCA products and what are the most effective ways to increase awareness? How, if at all, did involvement in the development process contribute to general awareness and use of the report?	
<ul><li>How did you hear about the NCA?</li><li>Are there specific actions you would recommend the NCA take to raise awareness of its products within your field?</li><li>If you do use NCA products, how do you stay in touch with new developments or releases?</li><li>Has your organization shared NCA products in any of the following ways (e.g., posted on website, shared in a newsletter, presented in a webinar)?<ul><li>*Why did you share what you shared? With whom did you share it? What feedback, if any, did you receive? Why did you decide not to share other USGCRP resources?*</li></ul></li><li>What resources do you find most valuable for learning about climate change in the United States and its impacts?</li></ul>	Secondary data analysis:<ul><li>Analysis of social media networks that observes how NCA products are propagated through them.</li></ul>Primary data analysis:<ul><li>Network developed through interviews/surveys with selected audience and snowball sampling to trace the network.</li></ul>
Evaluation Question 2: How, and to what extent, did NCA products address information needs among priority audiences (i.e., what did they gain cognitively in terms of knowledge, skills, attitudes, capacities, etc.)? **Evaluation Question 3: How, and to what extent, did NCA products address decision needs among priority audiences (i.e., what did they do as a result of using the products)?**	
<ul><li>To what extent do you find the NCA helped you to:<ul><li>Understand causes of climate change</li><li>Understand impacts of climate change</li><li>Learn about climate data</li><li>Become aware of a network of peers, resources, experts</li><li>Understand what you can do to address climate change</li><li>Understand solutions others can pursue to address climate change</li></ul></li><li>For each USGCRP product, indicate whether that product helped you to:<ul><li>Justify or affirm a decision or action</li><li>Inform an action or decision</li><li>Participate in a network</li><li>Seek additional information</li><li>Engage in other NCA activities</li><li>Share and adapt information with others</li><li>Identify research needs and gaps</li></ul></li><li>Can you provide examples of specific ways in which NCA products have helped you in your work?<ul><li>*For each example, tell us more about: (a) how you used it, (b) what specific parts were most helpful and why, (c) what was not helpful, and (d) what was the result?*</li></ul></li><li>Which NCA products would you like to see but currently do not exist?<ul><li>*What might you and your organization do differently if you had access to that type of resource?*</li></ul></li></ul>	Secondary data analysis:<ul><li>Review of policy documents, conferences, websites of organizations, and events relevant to the climate field.</li><li>Content-based citation analysis may be used to identify what NCA products are used in decisions; bibliographic coupling citation analysis may be used to identify how distinctive audiences use NCA products differently.</li><li>These analyses may also be used to identify subjects for case studies, interviews, or focus groups.</li></ul>

TABLE 6-2 Continued

Primary Data Collection: Examples of Questions for Surveys, Interviews, Focus Groups	Considerations for Analysis of Both Primary and Secondary Data
Evaluation Question 4: How did the attributes of the products and process contribute to how users feel, what they gain (e.g., cognition), and what they do (e.g., behaviors)? What about the products and process could be changed to make them more effective?	
<ul><li>On a scale of 1 to 10, how satisfied are you with NCA products overall?</li><li>To what extent do you agree or disagree with the following statements:<ul><li>The NCA covered my needs</li><li>The NCA is relevant to me</li><li>The NCA helped me do my job</li><li>The NCA is easy to access</li><li>The NCA is easy to use</li><li>The NCA is easy to share</li><li>The NCA is trustworthy</li></ul></li><li>On a scale of 1 to 10, how satisfied are you with specific aspects of USGCRP products (e.g., the selection of topics, the regional descriptions, NCA Atlas)?<ul><li>Describe how these various aspects of USGCRP products affected how your organization was able to use them.</li></ul></li><li>How could they be improved?</li><li>Which practices could make use of NCA products easier for you?</li><li>*How do NCA products compare with other products or sources you have prior experience with?*</li></ul>	Secondary data analysis<ul><li>Social media network analysis, records of meeting attendance, and citation analysis might detect different patterns of use of NCA based on whether individuals participated in the development process or not</li></ul>
Evaluation Question 5: How do the contextual factors described in the logic model influence how audiences feel, what they gain (e.g., cognition), what they do (e.g., behaviors), and how they mediate the use of the NCA?	
<ul><li>To what extent do NCA products address issues salient to equity?</li><li>To what extent to you agree or disagree with the following statements:<ul><li>Others like me use the NCA</li><li>People like me were included in the NCA process</li></ul></li><li>*What emerging needs related to climate change do you think USGCRP could address?*</li><li>*What sources do you trust most with regard to climate change?*</li><li>*Where do you usually go for information on climate change?*</li></ul>	Secondary data analysis<ul><li>A redundancy analysis can be used on a citation network to determine what the effects of removing USGCRP products would be.</li></ul>Primary data analysis<ul><li>Analysis of survey data stratified by respondent demographics can reveal if, for example, perceptions of how respondents feel about the NCA vary by demographic characteristic</li></ul>
Evaluation Question 6: How does the network of networks factor into use?	
<ul><li>*How do NCA products complement other resources you use concerning climate change and health?*</li><li>Are there particular influencers or central actors in your network that you think could help disseminate awareness about NCA use? If so, please list and elaborate.</li><li>*Is there anything else you would like to share about your experience with NCA products that you feel is important for the evaluator to know but was not asked?*</li></ul>	Secondary data analysis<ul><li>A multilayer network analysis based on citations could be used to trace how information flows between agencies, USGCRP products, and other audiences.</li></ul>

NOTES: NCA = National Climate Assessment; USGCRP = U.S. Global Change Research Program.

Also note that questions in surveys are more likely to have close-ended response options than those used in interview and focus group guides. To simplify matters, examples of questions that might be used for all three types of protocols are included below. Questions that are particularly suited for interviews or focus groups are included in italics. However, the evaluation team would refine these questions and, as described in Chapter 3, may engage in cognitive interviewing or other processes to fine-tune the questions and increase the likelihood that they will yield information in a manner that will benefit the evaluation.

The right column provides some considerations about analysis of data—either analysis of data collected through the evaluation itself (primary data) or analysis of secondary sources. It is also designed to provide examples of potential methods that are appropriate for the overarching evaluation questions, rather than a complete list of potential analytic tools.

ILLUSTRATIVE APPLICATIONS TO PARTICULAR AUDIENCES

This section considers five examples that illustrate how USGCRP can use a logic model, combined with the overarching goals of an evaluation and evaluation questions, to plan for data collection and other information-gathering activities from different types of audiences. These examples were selected to demonstrate how a diversity of users can be taken into account in designing an evaluation, as well as how the choice of evaluation methods can be tailored to not only the specific audience of interest but also the specific goal of the evaluation as it relates to use of the NCA by each audience. The examples are (1) federal agencies or programs, such as the U.S. Department of Agriculture (USDA) climate hubs, which collectively constitute a major audience for the NCA and also function as important nodes in a network of users; (2) community groups and nongovernmental organizations (NGOs), which can also provide nodes to networks that touch audiences that might be difficult or impossible for USGCRP to reach directly; (3) public health professionals, who serve a wide range of roles and interact with individuals and communities on many levels (from very local to national); (4) mass and social media, which form a widespread and diverse network focused on information sharing; and (5) K–12 educators, who reach audiences of important future decision-makers but are difficult to access directly. The purpose of the discussion here is not to imply that these five groups should be the priority audiences for inclusion in an evaluation of the NCA—that prioritization needs to be determined by USGCRP following criteria such as those outlined in Chapter 5. Rather, it is to illustrate the diversity of pathways through which audiences might make use of the NCA and its products, and how the Program needs to identify those pathways and associated key evaluation goals in order to choose appropriate methods and design plans for data collection.

Federal Agencies

Federal agencies both contribute to the NCA and provide important immediate nodes in the network of networks through which information in the NCA is shared. Many federal programs provide climate services and make NCA information and products available to regional, state, local, tribal, and other collaborators. Each program has an institutional commitment to a set of users and user pathways: each is designed to pursue and to facilitate uses of climate information, including that provided via the NCA. Examples of important federal programs that serve as both users and potential nodes in sharing information about or from the NCA include:

- National Oceanic and Atmospheric Administration's (NOAA's) Climate Adaptation Partnerships/Regional Integrated Sciences and Assessments program
- NOAA's National Centers for Environmental Information Regional Climate Center Program
- USDA's climate hubs
- U.S. Geological Survey's climate adaptation science centers
- U.S. Fish and Wildlife Service national landscape conservation cooperatives (LCCs)
- U.S. Department of Energy's Grid Resilience Technical Assistance Consortium[1]
- U.S. Department of Health and Human Services's (HHS's) Office of Climate Change and Health Equity (OCCHE)

---

[1] See https://www.energy.gov/gdo/grid-resilience-technical-assistance-consortium.

USDA's Climate Hubs

USDA's (n.d.-c) climate hubs illustrate how the framework developed in Chapters 3 and 4 could be applied to federal agencies' involvement in the NCA. The 11 hubs anchor collaborations "within their regions, which include scientists, practitioners, and local decision-makers, to identify regional vulnerabilities, scientific needs and data, and methods for mitigating impacts and adapting to changing conditions" (U.S. Climate Resilience Toolkit, n.d., p. 1). (See Appendix F for more detail regarding the operations and services provided by the climate hubs.) Each hub is tasked with addressing the unique climate challenges and opportunities of its region, while also working with other hubs to develop and share information nationwide (USDA, 2021). The USDA hub employees combine the job of researcher with the job of agricultural extension agent, forming a key link between the NCA and the agricultural, ranching, and forest management communities. They provide periodic regional assessments of risk and vulnerability to production sectors and rural economies, building on material provided through the NCA. USDA hub employees also serve as contributors to the NCA.

Additionally, USDA climate hubs collaborate with other federal agencies and work together with state, local, and tribal governments, as well as with universities. Together, the federal network and their collaborators at a regional level constitute a network of networks addressing "actionable scientific information that enables local decision-makers to work collaboratively, across landscapes, to implement adaptation strategies that reduce risk and build resilience" (U.S. Climate Resilience Toolkit, n.d., p. 2). The hubs form central nodes in a network that includes the following (USDA, 2021):

- Internal partners: USDA research agencies and program agencies.
- External collaborators: Other federal agencies, universities and extension, states, tribes, local governments, nongovernmental organizations, international organizations, and other technology transfer providers, including but not limited to natural resource managers and service providers, and regional boundary organizations.
- USDA customers: Farmers, ranchers, and forest managers, and the people and communities that depend on them. (p. 10)

Considerations for USGCRP Evaluations

To evaluate the use of NCA by federal agencies, using the climate hubs as an example, a first step is for USGCRP to clearly identify the goals of the evaluation and what they hope to learn from it (see Chapter 3), because this will inform not only the specific questions to ask but also the approach or methodology to use for data collection. For example, if the goal is to learn the extent to which there is familiarity with or use of NCA products within the hubs, then USGCRP could conduct a survey across all hubs, asking specific questions such as those described in the left column (data collection questions) of Table 6-2, above. Since the hubs are a well-defined, easily identifiable, and relatively small audience (11 hubs), a census (sent to all hubs rather than a representative sample) is feasible. The questions could be structured to collect quantitative data using, for example, prepopulated lists of potential responses rather than open-ended questions. The survey results could then be collated, summarized numerically, and reported.

Alternatively, if USGCRP wants to understand *how* the hubs are using the NCA and its products and how they might use them more effectively, then it might want to select a diverse subset of the hubs and conduct in-depth case studies of each, including interviews of key personnel, focus groups, etc. In this case, the interview questions could be more open-ended, asking, for example, where hub staff get their climate-related information, what they do to transform it, and what their dissemination strategies are. This would yield qualitative data that provide a more in-depth (but potentially less easily numerically summarized) picture of NCA use by the hubs.

Given the important role that the hubs play in sharing climate-related information and collaborating with other groups, the evaluation could map out the key networks through which the hubs operate and serve as a node.

This mapping accounts for transmission that is both vertical (to and from senior leadership) and horizontal (to and from federal and other collaborators, as well as outward to customers). After identifying these networks, information could then be gathered on the extent and ways in which the hubs use and share information about the NCA through these networks. Interviews, focus groups, and case studies are likely to be more effective than surveys in generating an in-depth understanding of the ways in which the hubs facilitate use of NCA-generated information by members in their network, particularly those members who are not easily identifiable by the evaluator and may not even be aware of the source of the information received through the hubs.

Finally, a practical consideration is that evaluation instruments can be used with federal employees without clearance procedures by the Office of Management and Budget (OMB), which are required by the Paperwork Reduction Act. While this requirement can be addressed, the evaluator and USGCRP should consider the appropriate times and places when focusing on only federal employees may offer tactical advantages.

Quick Summary: USDA Climate Hubs

Role in network of networks: information providers—they have internal partners within USDA; external partners including federal agencies, universities, states, tribes, and local governments; and USDA customers (e.g., farmers, ranchers, and forest managers)

Uses of NCA (actual or potential): translate and deliver NCA information (coupled with other information sources)

Membership: 11 hubs, well defined

Accessibility for evaluation: high—membership lists exist, OMB clearance not needed

---

Community Groups and NGOs

Many community groups value information about climate change. As used here, the term *community* is construed broadly. It can be place-based and unique to a particular group of people that lives there (Fitzgerald, 2018) or apply to individuals who share a common religion, core identity, or objective. Community boundaries can be perceptual (such as an ethnic identity) or actual (such as living in a specific neighborhood) and are frequently defined by inclusion and exclusion criteria (Fitzgerald, 2018). Clearly, different communities will have different informational and other needs (as reflected in the illustrative logic model in Chapter 3).

Communities may be direct users of the NCA and its products, but may also be indirect users through, for example, materials generated by NGOs. Many NGOs (e.g., 501(c)(3) or (4) organizations, public charities) represent or support community members by advocating for policy changes, offering services, raising awareness, and/or mobilizing resources to spearhead localized change. These entities can use the localized data from NCA products to inform their communities about the potential impacts of climate change, possibly via derivative products that make use of the NCA without the target audience necessarily being aware of the reliance on NCA information.

Union of Concerned Scientists

The Union of Concerned Scientists (UCS) is an example of an NGO that acts as an intermediary between the NCA and the general public. UCS is a nonprofit organization created to put science into action. Climate change is a priority topic for the group, and the NCA is one of their major information sources, along with National Weather Service data and alerts concerning heat, fire, floods, and storms. As an intermediary between the NCA and the public, UCS tailors and translates NCA information to fit local situations. For instance, UCS creates maps as an outreach tool to show how local communities are affected by climate change and how corresponding inequities arise, both directly through effects of the weather and indirectly such as through impacts on the energy grid. Through these efforts, UCS can provide localized assessments that go well beyond what is available directly in the NCA.

Considerations for USGCRP Evaluations

Given the broad definition of relevant communities, identifying the full set of community groups that might make use of the NCA for inclusion in an evaluation would be infeasible. Rather, USGCRP can engage a subset of groups in an evaluation (through, e.g., in-depth interviews or case studies) to appraise the success of the NCA process in understanding their needs and co-developing products tailored to meeting those needs. Under this approach, USGCRP could identify communities of high priority based on its logic model and what it wants to learn from the evaluation. The selection of communities could include, for example, communities where the effects of climate change are significant and where information of the type provided in the NCA might be particularly useful. It could also consider diversity, equity, and inclusion goals in selecting communities, particularly those that might face barriers to awareness, access, or use of the NCA and its products.

Alternatively, USGCRP could focus on evaluating indirect use by focusing on NGOs that make use of the NCA in preparing derivative materials tailored to specific community needs. This is likely to be a more identifiable audience. If USGCRP wants to see how widespread the awareness and use of the NCA are among NGOs and the extent to which they share information about the NCA, it could conduct a survey of major environmental nonprofit organizations such as UCS, The Nature Conservancy, Sierra Club, Environmental Defense Fund, and the Natural Resources Defense Council. It could also look at NGO websites to see the extent to which those organizations are utilizing knowledge from the NCA. If, on the other hand, USGCRP wants to learn about how NGO engagement with local communities impacts local decision-making, a case study of an organization such as UCS that works closely with local communities and provides localized information based on the NCA could be conducted to gain insight into how NCA information is translated for use at the local level and how future assessments and related products could facilitate this bridging function.

Quick Summary: Community Groups and NGOs

Role in network of networks: varied—they may create and disseminate information and/or be actively involved in addressing climate change, often at the local level; community groups may be users only

Uses of NCA (actual or potential): translate and deliver NCA information (and possibly other data sources); use of NCA information for planning and decision-making

Membership: an estimated 1.5 million NGOs working in the United States; the number involved in climate change activities is unknown; the number of community groups is unknown

Accessibility for evaluation: one might examine one or a few selected NGOs and/or community groups

---

Public Health Professionals

Public health is a policy arena that encompasses a wide range of occupations, united by concern for the health of human populations. This section discusses how the NCA is responding to public health concerns and illustrates methods that could be brought to bear in evaluating the outcomes of the NCA on this heterogeneous audience. Moreover, this example can be extrapolated to the many similar professional networks that intersect with climate impacts across a range of disciplines.

Public Health Professionals and Climate Science

The relevance of USGCRP to public health is reflected in the many parts of the fifth NCA that specifically address issues salient to public health research and practice, including Chapter 15 (Human Health), Chapter 31 (Adaptation), Chapter F3 (COVID-19 and Climate Change), and mentions of health threats in other topically and regionally focused chapters. Public health professionals are integral to the identification of climate impacts on public health (Frumkin et al., 2008; Romanello et al., 2023), estimation of health co-benefits of climate action (Roca-Barceló et al., 2024), advocacy for climate change mitigation (Kreslake et al. 2018), and evaluation of climate adaptation efforts (Joseph et al., 2023; Turek-Hankins et al., 2021). They hold roles such as science communicators, researchers, governmental staff, and employees of NGOs and private-sector companies; they work across multiple sectors and can be found in academic, state government, nonprofit, and research institutions. In these roles, they might reciprocally contribute evidence that informs USGCRP and refer to the NCA and other products to contextualize their work (USGCRP, 2016a). In addition, federal scientists work in agencies with a public health mandate (e.g., U.S. Centers for Disease Control and Prevention, HHS's OCCHE) (USGCRP, n.d.), which may be major users of USGCRP products and may produce secondary products that influence public health practitioners in state, tribal, territorial, and local agencies.

Public health practitioners working in state, tribal, territorial, and local agencies also play critical roles in tracking population health outcomes and supporting public outreach and engagement on climate adaptation measure implementation (Albright et al., 2020; Errett et al., 2022; Frumkin et al., 2008; Kreslake et al., 2018; Marinucci et al., 2014; Romanello et al., 2023; Rudolph and Gould, 2015; Sheehan et al., 2017). These professionals may be consumers of the NCA and related products as a credible source of information on which to base outreach efforts. Other resources, including those compiled by HHS's OCCHE, are also oriented toward community public health practitioners and might be tapped in an evaluation of the NCA and related products.

Considerations for USGCRP Evaluations

Public health professionals constitute a diverse community of practice, and the degree of direct and indirect use of NCA products and information will vary with the individual's role. For example, the perceptions surrounding NCA products and their effects on knowledge and capabilities and on decisions and practice may differ according to the type of public health worker (e.g., county agency staff, state research scientist, academic researcher). Evaluation of the use of the NCA and its related products by this audience would require that the evaluation questions and methods be tailored to a subgroup that has been prioritized using the criteria in Chapter 5 and based on the logic model (see Chapter 3) and the overall goals of the evaluation.

For example, if USGCRP wants to understand use of the NCA by local public health agency professionals, they might conduct a survey that includes questions from Table 6-2 about norms (e.g., "Is this what others like me are using?"), efficacy beliefs (e.g., "Does this help me to do my job?"), and control beliefs (e.g., "Is this easy to access?"). Those norms and beliefs might influence the probability that the individual would use a USGCRP product to "learn about climate data," "understand solutions and what can be done by myself and others," or "access and use climate data." As an alternative to conducting its own survey, USGCRP could seek to add questions to surveys already being conducted of those professionals such as the National Association of County and City Health Officials' Forces of Change survey[2] or the Public Health Workforce Information and Needs Survey.[3]

In addition, since the public health community has multiple sources of information on climate and health, including the CDC, fact sheets produced by the American Public Health Association (APHA), and information provided by OCCHE, it would also benefit an evaluator to engage with some of these agencies' staff to learn about whether and how they use NCA products to inform materials for local public health practitioners. It is also important for the evaluator to recognize that there may be differences in perceptions about norms and accessibility between USGCRP original products and translational secondary products (e.g., fact sheets).

If USGCRP's goal is instead to understand how CDC researchers use the NCA in their research, a different set of questions would be needed. For example, questions may focus on whether the data provided by USGCRP (e.g., in the NCA Atlas) are granular enough and formatted in such a way to allow for easy analysis. USGCRP could also conduct citation analyses of CDC publications—both scientific publications (e.g., the *Morbidity and Mortality Weekly Report*[4]) and those designed for a broader audience, such as newsletters—to see the extent to which they reference the NCA.

---

[2] See https://www.naccho.org/resources/lhd-research/forces-of-change.
[3] See https://debeaumont.org/phwins/what-is-phwins.
[4] See https://www.cdc.gov/mmwr/index.html.

> **Quick Summary: Public Health Professionals**
>
> **Role in network of networks:** varied—they may be more likely to be users of NCA information than creators or disseminators of climate change information
>
> **Uses of NCA (actual or potential):** translate and deliver NCA information (and possibly other data sources); use NCA information for planning and decision-making
>
> **Membership:** categories may include epidemiologists, public health nurses, health educators, environmental health specialists, health administrators/managers, biostatisticians, and health policy analysts; they are highly dispersed—APHA estimates there are a minimum of 500,000 in the public health workforce (Perlino, 2006)
>
> **Accessibility for evaluation:** complete membership lists not available; might start by looking at a single public health organization or to look at a more narrowly prescribed subset of public health professionals (e.g., local health department officials)

Media

The mainstream media is a network through which information from the NCA and its products is distributed. This broadly includes entities that generate and distribute content via cable, print, broadcast, and digital television and those that use contemporary media platforms such as social media, short-form media, and the worldwide web. Decision-makers and the American public follow the news, and the news that the media choose to cover or not cover often sets the social agenda (Simons, 2021). Some news outlets report on a broad range of topics, while others focus on specific topics. For example, Climate Central is a nonprofit news organization that both conducts research and reports on climate science and the impacts of climate change.

Media as a Network of Networks

Furthermore, media operate through a network of networks, as they continue to influence public life. A news article could report an NCA finding, which would then be discussed and reported on online platforms (McNamara, 2009). For example, a television reporter or journalist could turn to the NCA to obtain information on climate change or to use data to create localized footage for their specific television or print market, which could in turn be picked up by other media. Similarly, media could pass on institutional knowledge and best practices regarding the value of the NCA to their peers, further affecting the dissemination of NCA products.

The news media face daunting obstacles in spreading awareness and accurate information about topics of interest to the public. These include the closure of regional newspapers (Gorman, 2011); a highly politicized atmosphere (Bykov, 2021) in which people turn to the media to reinforce their preexisting views; and the proliferation of misinformation, disinformation, and malinformation (Hunter, 2023).[5] Climate misinformation and its spread through the media, particularly social media, is an issue of particular concern. For example, in a review article, researchers noted that there appear to be organized networks supporting climate misinformation and that these networks are highly organized in the production and dissemination of misinformation (Treen et al., 2020; Turrentine, 2022). An NCA evaluation could shed light on what role, if any, NCA information plays in these misinformation networks and how these practices could be countered. Moreover, scholars have also demonstrated that strategies for countering disinformation encompass, among other things, educating the public about disinformation and "innoculating" them by disseminating examples of disinformation (Roozenbeek et al., 2020). Evaluators therefore need to be aware of the possibility of misinformation and disinformation

---

[5] Information may be problematic because it is false but with no intention of causing harm (misinformation), false with the intention to create harm (disinformation), or based on fact but used out of context to mislead, harm, or manipulate.

when evaluating NCA use and consider measures to assess whether or not users have made accommodations to address this problem.

As an adjunct to the mainstream media, use of social media has been argued to serve as a way to overcome the power of media elites and make media distribution, curation, and creation more accessible and democratic (Anstead, 2015). The climate youth movement and its sister movements have relied heavily on social media to spread their messages and rally people to their cause (Fisher, 2012). The spread and use of social media, particularly by younger audiences, represents an opportunity to enhance awareness and use of NCA products—however, social media can also contribute to misinformation, disinformation, and malinformation (Van der Linder et al., 2017).

Considerations for USGCRP Evaluations

To learn how extensively the media contribute to sharing information about the NCA and its products, USGCRP could use media analysis, including, for example, a scraping of newspaper stories or social media posts for mention of NCA. As part of a process of ongoing evaluation, USGCRP could also subscribe to media-tracking software such as TVEyes Media Monitoring for ongoing use. Such data can provide measures of the extent and nature of the sharing of information about the NCA throughout the media network, and the types of individuals and organizations that are reporting on the NCA and its products. A focus on explicit mentions would not include all the ways in which NCA influences the media, however, as USGCRP products are also used indirectly (with or without attribution) in media products related to climate change.

An evaluation of how the media are using the NCA could include surveys, interviews, or focus groups of journalists, as well as search tools such as LexisNexis. Science journalists represent one audience within the broader media audience that may rely on NCA products to help them carry out their duties. Identifiable audiences include the Society of Environmental Journalists and Internews' Earth Journalism Network. Other groups, such as the Society of Professional Journalists or the National Association of Hispanic Journalists could be included if USGCRP wants to learn about coverage of climate change and climate science.

Alternatively, to understand more fully how the media are using (or not using) the NCA, USGCRP could do an in-depth case study of use by an organization such as Climate Central. Such a case study could draw on interviews with staff (scientists and journalists), as well as website and citation analysis.

Quick Summary: Media

Role in network of networks: their primary role is to disseminate information

Uses of NCA (actual or potential): deliver information from the NCA and other sources

Membership: 3,000 newsroom outlets, plus blogs, podcasts, talk radio, and other digital content

Accessibility for evaluation: could be examined through media search tools or a case study

---

K–12 Educators

One audience of significance is the population of the future, which is likely to be exposed to increasingly severe impacts attributable to a changing climate. A portion of that audience is today's students, whose experience of the NCA can be evaluated via their teachers. The challenge of including educators in an evaluation of the NCA illustrates some of the difficulty of understanding how climate science reaches, and fails to reach, the general public.

Increasingly, K–12 educators are teaching their students about climate change, including how the climate is changing and how those changes impact both their own lives and the lives of others, now and into the future. Some educators also emphasize resilience strategies, to support students who may be impacted by extreme weather events (Dubois and Krasny, 2016).

The NCA and its associated products can support educators and students in two ways. The first is by providing information to those educators so that they can learn about climate change and use that knowledge in their educational mission. For this, educators might access the NCA directly or might learn about it and the information in it through another audience, as part of a network. The second is by providing resources that can be used in the classroom to teach students about climate change in a way that will be effective and engaging. In all areas, including climate change, educators face the task of communicating complex information to students in ways that are tailored to students' needs (Kirk et al., 2014; Monroe et al., 2019), which scholars have demonstrated is crucial if the intention is for the audience to understand and act upon the message (Cummings, 2017; Nisbet, 2009). Moreover, as busy professionals (Walker, 2023) for whom climate science is not a primary mission topic, K–12 educators are likely looking for easy-to-share, easy-to-understand, and trustworthy information they can use in their classrooms.

According to a review article analyzing 959 studies on climate education interventions, successful environmental education had two things in common: (1) content was personalized and made specifically relevant to the student, and (2) teachers used active, engaged learning while working hands-on with students (Monroe et al., 2019). Educators seek trustworthy information to introduce into the classroom and often rely on established sources, including listings on the internet (Puttick and Talks, 2021).

Educators and the NCA

Some NCA-related tools that could be particularly useful to educators in the classroom include the U.S. Climate Resilience Toolkit[6] and the NCA Interactive Atlas.[7] For example, students may be motivated to learn more about the communities they care about by seeing how farmers are adapting their growing methods to cultivate strawberries in Florida's changing climate or how the urban heat island effect is influencing Chicago.

Researchers have found that teachers often rely on four types of information sources when it comes to educating students on climate change: the internet, government sources, the media, and professional development courses, according to a scoping review of nearly 600 papers (Puttick and Talks, 2021). Sites that aggregate listings for teachers, such as commonsense.org, may be worth investigating, since educators use them to identify trustworthy sources for teaching about climate. This website, for example, lists National Geographic Education and NASA Global Climate Change—Vital Signs of the Planet as top sources, but does not reference the NCA.

Considerations for USGCRP Evaluations

USGCRP could evaluate the use of the NCA by K–12 educators through a network of networks approach, first looking at whether there are particular educator nodes that the NCA is or is not collaborating with to disseminate awareness of NCA or that are being used to shape derivative products. Possible key educator nodes include professional organizations for teachers and K–12 leaders, Ted-Ed, Scholastic Education, PBS Learning Media, and Open Educational Resource Commons. Within these nodes awareness of specific tools should be evaluated (see Behl et al., 2015) using focus groups, surveys, or one-on-one interviews. Given the importance of education for providing underserved populations with climate information, an evaluation might also identify target subpopulations based on diversity, equity, inclusion, and justice considerations, to determine, for example, whether NCA products or tools are being used in ways that improve resilience of economically marginalized groups.

USGCRP could also collect information on the use of the NCA by reviewing curricula and programming for references to the NCA or NCA products. Sources for review could include resources listed on commonsense.org or the climate change resources provided by the National Science Teachers Association.

---

[6] See https://toolkit.climate.gov/content/us-climate-resilience-toolkit.
[7] See https://atlas.globalchange.gov.

> **Quick Summary: K–12 educators**
>
> **Role in network of networks:** their primary role is to disseminate information
>
> **Uses of NCA (actual or potential):** probably through intermediary sources
>
> **Membership:** 3.8 million full- and part-time public school teachers
>
> **Accessibility for evaluation:** no list exists; could be surveyed through two-stage sample (schools first, and teachers within schools); could review curricula or look at educator nodes

CONCLUSIONS AND RECOMMENDATIONS

Conclusion 6-1: A variety of methods is needed in an evaluation of the NCA; some of these collect and analyze primary data while others are used to analyze existing data. Each has advantages and disadvantages, and they vary in the extent to which they address different types of questions. Surveys are well suited to gathering some information from a larger number of individuals and focus groups, while in-depth interviews, case studies, and observations can provide more nuanced understanding of how climate science information is used and how it is disseminated through networks.

Conclusion 6-2: Audiences make use of the NCA and its products through diverse pathways. Identifying those pathways and associated key evaluation goals is needed in order to choose appropriate methods and to design plans for data collection.

Conclusion 6-3: Audiences will differ in their characteristics, including how easily they can be identified and contacted for data collection. This will impact the feasibility of using different methods to collect primary data and availability of existing (secondary) data.

Recommendation 6-1: The U.S. Global Change Research Program should design its evaluation and data collection plan so that the methods used for priority audiences can answer the overall evaluation questions identified in the logic model. The methods and approach chosen should be tailored to the audience and the evaluation question being investigated.

7

Implementing the Evaluation for Continuous Improvement

The scope of an evaluation of the kind described in Chapters 3–6 is ambitious because the mandate of the U.S. Global Change Research Program (USGCRP) is ambitious: climate change affects everyone and all sectors of the economy. Over time, USGCRP has broadened the range and number of experts and interests that it engages in developing the National Climate Assessment (NCA), and the assessment itself has grown. As a result, carrying out an evaluation of the NCA and associated products is a challenging task. One overarching consideration is the need to continuously learn from the results of evaluations, incorporating knowledge gained into future efforts. This is important for a dynamic program such as USGCRP with large-scale, complex audiences. USGCRP's work is nested within expansive networks of activity and the USGCRP is an important convener of activity within those networks. Part of the value of incorporating a network perspective into evaluation work is that it embraces the dynamic nature of how this web of relationships evolves over time. Who is included or left out, events or entities emerging and disappearing, structural holes opening and closing are all examples of how the context USGCRP is operating within can change rapidly—especially when USGCRP itself intervenes in these networks. How information disseminates and the impact of strategies for reaching targeted audiences may change significantly over time.

The NCA itself also continues to change, producing new aspects to examine with every iteration. The cyclical nature of many USGCRP processes and products offers opportunities to measure whether changes made in response to an evaluation have resulted in improvement toward goals, and to evaluate the success of process innovations as well as product outcomes. Careful attention to the timing and sequencing of evaluation efforts will result in more impactful products, and each phase of an evaluation strategy can be informed by the knowledge that has been gained in previous phases.

TIMING AND SEQUENCING OF EVALUATION WORK

Because the work of USGCRP is continuous, but the NCA and specific products are released at particular times, attention to timing and sequencing are important to designing an evaluation, with important considerations related to the sequences for both development and evaluation of the NCA, needs of evaluation users, and the time lag between release of an NCA and the decisions and actions that may be taken based on that assessment.

Leverage the NCA development process in the evaluation. At any point in time, USGCRP is simultaneously disseminating an existing NCA and developing the next. Evaluators should think about how an evaluation design can both inform and leverage the NCA process. For example, NCA-related workshops and engagement activities could be avenues for evaluators to conduct observation or focus groups to gather information on the use, utility, and gaps of prior NCAs as well as desires for future improvements.

Consider the needs of evaluation users. In order to support continuous improvement of USGCRP products, one goal could be to release evaluation findings before key milestones, for example, before authors begin outlining NCA chapters or before a concept for the next generation of NCA tools is finalized. To achieve this, it would be important for evaluators, USGCRP, and key evaluation users to identify when evaluation outputs could inform decision-making during the NCA process.

Consider the lag between the release of one NCA and the outcomes anticipated in the logic model. Decisions that incorporate the NCA and other USGCRP products may not be apparent or measurable for some time after the release of a product. One approach to addressing this time-lag issue is to plan ongoing or sequenced evaluation activities to capture information about activities soon after they are concluded, as well as longer-term or ripple effects.

Sequence the evaluation. Planning and resourcing a single-stage evaluation is impractical, if only because the scope of uses and the range of audiences can only be uncovered in stages. Evaluating the uses of evolving USGCRP products implies the need to continuously learn from previous efforts, and some information can only be collected in steps, based on what has been learned previously.

Several methods may help sequence evaluation processes:

1. Conducting an *evaluability assessment*, an approach to determining how feasible it is to conduct an evaluation that can help improve evaluation design before full implementation begins (Trevisan, 2007). This could entail customizing a logic model for the product at hand; prioritizing USGCRP audiences; identifying the needs of evaluation users and their overarching evaluation questions; developing and testing data collection instruments; and then investigating what data sources are, or could be, available. The availability of data related to equity is an important consideration in evaluability assessments (Lam and Skinner, 2021). As such, the evaluability assessment may help determine the availability of sociodemographic data related to use of USGCRP products and may yield insights on how to understand who is and who is not using these products.
2. Piloting a network of networks assessment on one or two agencies. Comprehensively mapping the effects of USGCRP products through each of the participating government entities would be a significant effort; starting with a limited number is more feasible. A pilot would enable testing of the approach, provide understanding of what is entailed, and offer opportunities to refine methods before rolling it out across multiple agencies. It would be important to be thoughtful about the selection of agencies to focus on first and to recognize the operational and organizational differences among USGCRP agencies.
3. Focusing an initial evaluation on one or two chapters of the NCA. For example, the evaluators could apply the logic model, overarching evaluation questions, and specific data collection methods to better understand how the information presented in the NCA chapter on water or energy meet the information and decision needs of the priority audiences. This could provide valuable opportunities to refine the approach before evaluating the reach of all the chapters.
4. Pretesting data collection instruments before using them. Pretests help to verify that questions are properly understood, that the right options are offered for responding to the questions, and that the respondent is able to provide accurate answers. Pretests also provide a way to obtain some early data that can inform the full evaluation.
5. Collecting the low-hanging fruit first. Some data are relatively easy to collect, such as when the population is well defined or the data to be considered are readily available. In other cases, the group to be studied is amorphous, with no list from which to draw a sample, and/or the group is unlikely to be able to provide detailed data. Starting with the easy routes can provide short-term feedback while longer-term, ongoing evaluation is planned. (Audience prioritization is discussed more in Chapter 5.)

Seek opportunities to support ongoing assessment. As described in Chapter 2, there has long been an interest in developing an assessment process that "incorporates ongoing evaluation of effectiveness, which facilitates adaptive management; supports adaptation actions across various timescales; stimulates civic engagement; and enhances the nation's capacity to effectively respond to the many challenges of accelerating global change" (Buizer et al., 2013, p. 16; see also Moss et al., 2019).

Collecting feedback on a continuous basis can support process improvement (see Chapter 3). For example, USGCRP may wish to collect feedback on its workshops and meetings so it can quickly implement improvements. USGCRP may want to monitor other metrics on an ongoing basis, such as tracking which products or chapters are most frequently used, in order to focus dissemination strategies. Network analysis could be used to examine changes in the networks and how they are tied to the NCA; the potential for such changes could be built into the logic model, and over time the findings might result in changes to the logic model.

Nonfederal participants have played a major role in the NCA since the beginning, and as contributors they are by definition first-order nodes in the network of networks. An evaluation might focus on how these participants' use of the NCA affects how they contribute to the dissemination of information, as well as how they contribute to future NCAs. Case studies may be helpful in understanding how the dynamic and complicated collaborations among USGCRP, its federal partners, and nonfederal participants have evolved over time and how they can be improved going forward.

On the other hand, the type of outcome evaluation described in this report is not well suited to be implemented in a continuous manner. Outcome evaluations provide information on results and perceptions at a particular point in time. Yet, there are opportunities to use a more formal outcome evaluation (as described in this report) as a starting point for ongoing evaluation. For example, if an evaluation develops an effective methodology for conducting case studies about use, USGCRP could apply that methodology to develop future case studies. USGCRP could also develop less formal case examples through other methods, such as an online forum where audiences could provide brief examples of how they have used the NCA or other products. Similarly, an outcome evaluation might develop and refine survey instruments to elicit feedback about audience perceptions of and actions taken with the NCA. Of the questions that prove most valuable in the initial survey, a limited set could be selected to solicit ongoing feedback through tools such as online forms.

COMMUNICATION OF EVALUATION FINDINGS

Communication of evaluation findings is of instrumental importance to ensuring that results will be understood and utilized in a way that informs policy and programmatic changes (Greene, 1988; Neuman et al., 2013). An evaluation effort is best designed in light of its intended use, and its communication strategies and products planned to include sufficient time to meet the needs of evaluation users, to receive feedback before finalizing findings, and to disseminate results effectively. As mentioned above, USGCRP is often simultaneously disseminating an existing NCA, developing related products, and planning the next NCA, meaning that ongoing evaluation and timely communication about results is necessary to inform continuous improvement.

The broader decision science and research communication literature points to a number of best practices in communicating evaluation results. One key is communicating with those involved in the evaluation about how their input was used, what was learned, and what is being done about it. Incorporating participants' perspectives can also inform prioritization of future efforts (Oliver et al., 2004, 2008; Rosenstock et al., 1998). Demonstrating that participants' input has been acknowledged, understood, and used can also help to improve the inclusivity of future efforts, helping to create a more participatory evaluation framework (Colorado Trust, 2002) to support ongoing learning and improvement.

LEGAL AND PROCESS CONSIDERATIONS

Any evaluation funded by the federal government will need to be cognizant of a range of applicable statutes and best practices, including Section 508 of the Rehabilitation Act[1] for inclusion and accessibility, the Freedom

---

[1] Rehabilitation Act of 1973, Public Law 93-112, 93rd Cong. (September 26, 1973), as amended through Public Law 117-286 (December 27, 2022).

of Information Act (FOIA),[2] the Federal Advisory Committee Act (FACA),[3] and others of which USGCRP staff and member agencies are well aware.

Note that the Paperwork Reduction Act (PRA)[4] places important limitations and restrictions on the demands federal agencies can place on the public to respond to information collection. Agencies' obligations under the PRA exist regardless of the method of collection—paper, virtual/online, or other survey mechanisms. Although expansive, there are some limitations to the law's application—for example, when fewer than 10 people are surveyed, or data are generated during discussion at a public event (online, hybrid, or in person). Neither is PRA approval required if federal employees are surveyed as part of federal work duties (see "When Doesn't the PRA Apply," GSA and OMB, n.d.). These exceptions point to the potential advantages of small, in-depth case studies with targeted, case study–specific questions to provide information about the variety of mechanisms in which the NCA is used in decision-making. It will be important for evaluation efforts to consider PRA requirements and, where the PRA applies, it will be necessary to factor in the time required to get clearance from the Office of Management and Budget (OMB).

USGCRP may also want to explore the possibility of engaging with evaluators that are not sponsored by the federal government. External evaluations could make important contributions to an ongoing process for evaluation and learning. There is a history of evaluations of the NCA and USGCRP conducted by independent academic researchers (Jacobs et al., 2016; Meyer, 2011; Morgan et al., 2005; Moser, 2005; Parson et al., 2003), but often based only on publicly available information. USGCRP could gain much more from such evaluations by establishing ways to collaborate, within the limits of law and policy, with interested researchers. Collaboration would allow the Program to communicate its priorities for evaluation and learning, as well as ensure that the researchers understand enough about the Program's inner workings to develop useful, actionable evaluation results. In addition, external evaluators (with nonfederal funding) have some freedoms that federally funded researchers would not, such as not being subject to the PRA or FACA. External evaluators who are given insight into the program could conduct an array of analyses that, with guidance on the Program's needs, could support ongoing learning, prioritization, and improvement.

CHOOSING A CONTRACTOR

It is likely that USGCRP will benefit by bringing in outside resources such as a contractor to assist with the evaluation. Performing an evaluation of this type can be a large undertaking, requiring a wide range of expertise, a substantial commitment of personnel time, and possibly specialized tools. USGCRP will still need to devote time to work with and monitor the contractor, but this would be much less of a stretch than trying to do the evaluation with Program staff alone. Following are some particular ways in which an outside contractor can be helpful:

- An outside contractor may offer a set of skills that USGCRP may not have had a need to acquire.
- Some contractors have prior experience with helping federal agencies obtain OMB clearance, which can be a difficult process for those who are unfamiliar with it.
- Some tasks can be labor-intensive, and USGCRP may not have sufficient staffing.
- Some contractors offer their own institutional review boards (IRBs).
- Some offer specialized software for sampling, data collection, and data analysis.
- Some offer facilities for large mailouts or telephone interviewing.
- Contractors have tools and personnel for handling the logistical work involved in an evaluation.
- There can be value in working with an independent entity, both for internal planning and for outside contacts. For example, some people might be candid with an outsider who has promised confidentiality, sharing information they would be reluctant to share with people with whom they work.

---

[2] FOIA Improvement Act of 2016, Public Law 114-185, 114th Cong, 2nd sess. (January 4, 2016), §552

[3] Federal Advisory Committee Act, Public Law 92-463, 92nd Cong. (October 6, 1972). Amended by Public Law 117-286, 117th Cong. (December 27, 2022).

[4] Paperwork Reduction Act, Public Law 104-13, 104th Cong. (May 22, 1995).

The term *contractor* here is used somewhat loosely, including, for example, contractual arrangements, grants, or even interagency agreements. It also might include a group of organizations, each providing a particular set of services; for example, there might be multiple contracts or a single contract that includes a prime contractor and a subcontractor. In such cases, it is important to clearly delineate the respective responsibilities and to have clear lines of authority and responsibility.

Specific requirements that might be specified in a request for proposals or grant announcement include the following:

- Successful experience in conducting evaluations of this type
- Expertise in network analysis, case studies, survey research, study design, logic models, cognitive interviews, web scraping, citation analysis, and mixed-methods research (i.e., quantitative and qualitative methods)
- Experience in preparing OMB packages and working with IRBs
- Sufficient number of staff to perform case studies, interviews, and surveys
- Appropriate software and physical facilities to support an evaluation

The indefinite nature of this work adds complexity in creating contracts of this type; the cost and level of effort may vary depending on which audiences are selected and what approaches are used to examine their use of the NCA. One approach could be designing a contract that has both base requirements (e.g., developing a logic model; creating an evaluation design; performing preliminary research to identify the audiences to be considered, including through the use of network analysis) and optional tasks (e.g., survey research, case studies) that might be awarded depending on the findings in the preliminary phase. Also, the optional tasks might be billed on a time-and-materials basis to allow for variations in size across the potential audiences.

CONCLUSIONS AND RECOMMENDATIONS

Conclusion 7-1: Careful attention to the timing and sequencing of evaluation activities is essential for realizing the potential for evaluation to inform future improvements to the NCA and its associated products.

Conclusion 7-2: Effective integration of communication strategies within this timing and sequencing of evaluation is critical for ensuring that the results are understood and used, and that they meet the needs of decision-makers.

Recommendation 7-1: In implementing evaluation for the National Climate Assessment and other products, the U.S. Global Change Research Program (USGCRP) should adopt a strategy that enables ongoing learning about how the processes and products are informing decisions, in order to support continuous improvement in USGCRP processes and resulting products.

Recommendation 7-2: The U.S. Global Change Research Program should sequence evaluation into manageable components, allowing for iterative testing and learning about how to best pursue evaluation over time. Sequenced components may include conducting evaluability assessments, piloting focused on certain agencies or chapters of the National Climate Assessment, picking low-hanging fruit first, or developing case studies.

Recommendation 7-3: In communication about evaluation efforts, the U.S. Global Change Research Program (USGCRP) should aim for active two-way communication with users. Communication mechanisms may include ongoing feedback, interim findings, meetings to tailor the communication of evaluation findings to particular situations, and communication about how input was used that helps connect evaluation efforts with USGCRP's objectives.

Recommendation 7-4: The U.S. Global Change Research Program should consider bringing in outside expertise and research capabilities—such as through contractors, consultants, grantees, or interagency agreements—to assist in designing and implementing the evaluation.

8

Putting It All Together

The National Climate Assessment (NCA) incorporates an extensive process to verify the scientific information contained within it. What has largely been missing, however, is an evaluation of how well the NCA works in terms of supporting those who have need for its information. The evaluation strategy discussed here is to meet that second goal and is directed at providing both measures of what the NCA has accomplished and information on how the NCA can be improved.

This report presents the key ingredients to be considered when the U.S. Global Change Research Program (USGCRP) designs evaluations, by offering at least a preliminary understanding of what might be learned through evaluation (overarching evaluation questions and logic model), who the Program wants to learn from (audiences), how the Program can learn from them (methods), and how to approach the process. Not everything can be done at once, and sometimes exploratory work will be needed to collect the information needed for later research. This chapter suggests a multistep approach to design, implement, and learn from evaluation, and offers a strategy for USGCRP to consider as it undertakes evaluation design.

ASKING THE RIGHT QUESTIONS

The answers that an evaluation can provide depend greatly on the questions asked. For this reason, the initial stage of designing evaluation questions is a key component of evaluation studies. Sometimes as an evaluation progresses, new questions are suggested and modifications can be appropriate. But because the initial stage is so important, the committee recommends creating a logic model to help generate the evaluation questions. Chapter 3 develops an illustrative logic model to serve as a foundation.

Often in program evaluations, one can identify highly specific outcomes of interest, such as an increase in graduation rates or a lowering of crime rates. The NCA is different in that its charge is to provide information, not to take specific actions with regard to climate change. One way of defining its goal is to say that the NCA exists so that decision-makers and policymakers have the information they need to make good decisions. That goal might be broken into three parts, as shown in Figure 8-1: (1) decision-makers need to have positive feelings about the NCA (e.g., they trust it and feel that it meets their needs), (2) they need to gain something from it (e.g., knowledge, connections, or solutions), and (3) they need to be able to act on it (e.g., through their decisions). At the same time, decision-makers operate in a larger context with multiple pressures to address climate change and multiple sources of information. By thinking about what each part involves and how those parts are connected and fit into a broader context in which there are multiple sources of information, one can develop a set of overarching

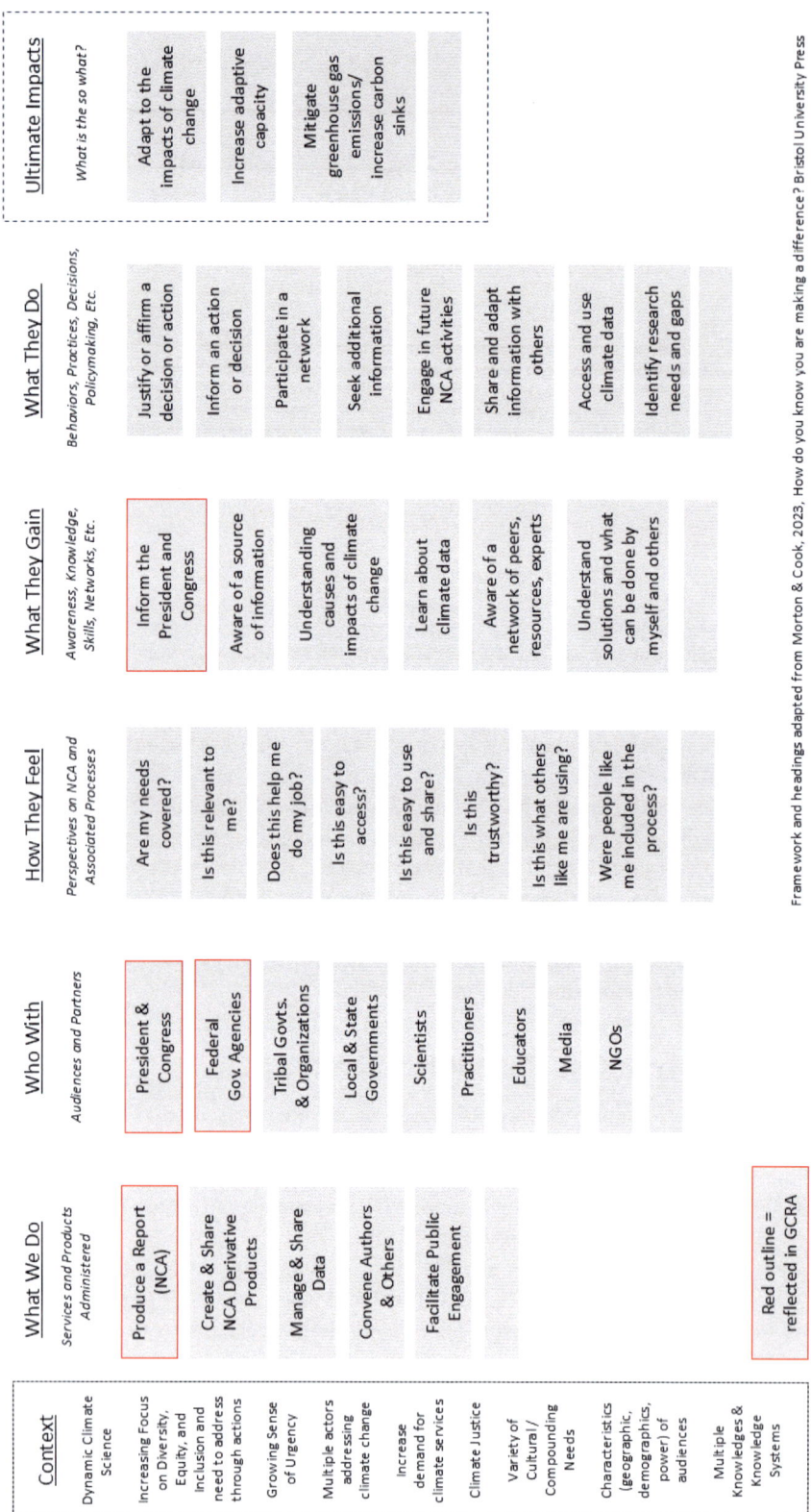

FIGURE 8-1 Illustrative logic model for an evaluation of the National Climate Assessment (NCA).
NOTE: GCRA = Global Change Research Act; NGO = nongovernmental organization.
SOURCE: Generated by the committee, adapted from Morton and Cook, 2023.

evaluation questions, as listed in Table 3-1. These overarching questions in turn can be adapted and made more specific to apply to particular contexts, and used, for example, in survey questionnaires or in-depth interviews. They also might be used through other types of methods, such as citation searches.

> **Recommendation 3-3: The U.S. Global Change Research Program should develop a logic model to describe how its products, including the National Climate Assessment, are hypothesized to achieve their intended outcomes.**

> **Recommendation 3-4: The U.S. Global Change Research Program should use a logic model developed for the evaluation to generate a set of overarching evaluation questions and should consult with partners and selected evaluation users to ascertain whether answering those evaluation questions will meet evaluation users' needs.**

DEALING WITH MULTIPLE AUDIENCES

A complicating factor, recognized in the Statement of Task, is that multiple audiences are dealing with and affected by climate change. Different audiences have different needs, different levels of information, different goals, and take different kinds of actions. A question that is appropriate to ask one audience may not be meaningful to another. Thus, any evaluation will require multiple data collections, and USGCRP will be faced with the task of deciding which audiences to focus on.

There is no single correct solution; rather, many solutions could work. Some audiences are so important that they need to be included (e.g., the legislation creating USGCRP specifies the president and Congress as the primary audiences of the NCA), and some audiences are less important in terms of making decisions related to climate change. What may be more important is how a set of audiences works as a collectivity—in other words, that each audience queried in the evaluation fills an important gap in knowledge of the NCA's outcomes.

One way to deal with multiple audiences is to perform network analysis—that is, to look at how different audiences are connected to each other and to the NCA. This provides insights into how information is disseminated and may indicate that certain audiences act as key nodes through which others are informed. Chapter 4 discusses some of the techniques that could be used. An advantage of this approach is that it provides a more comprehensive view of the entire network of networks concerned with climate change than would studies of individual audiences. It also may provide guidance as to which audiences are most suitable for more detailed examination.

> **Recommendation 4-1: In designing an evaluation of the National Climate Assessment or related products, the U.S. Global Change Research Program should make use of network analysis as a tool for addressing the evaluation questions related to understanding who key actors are, how information is transmitted across multiple entities, which entities serve as key nodes for disseminating information, and how the network of networks supports that flow of information.**

Another approach is to establish criteria for choosing audiences for more extensive study. In Chapter 5, the committee suggests two broad categories of criteria: those that meet USGCRP's information needs, and those are most feasible.

> **Recommendation 5-1: In choosing which groups to study as part of an evaluation, the U.S. Global Change Research Program should seek diversity (including a focus on marginalized populations) similar to that of the participants and audiences with which the Program seeks to engage.**

> **Recommendation 5-2: The U.S. Global Change Research Program (USGCRP) should select audiences to include in evaluation based on the following criteria: importance in USGCRP's logic model, including (1) the role of an audience in climate-related decision-making; (2) the role of an audience in the transmission of climate information for decision-making; (3) the generalizability of results from an**

audience to other populations; and (4) feasibility, diversity, and suitability for the evaluation question and method used. A targeted audience does not need to meet all of these criteria, but the audiences prioritized in an evaluation should meet these criteria collectively.

Recommendation 5-3: The U.S. Global Change Research Program (USGCRP) should continue to pay attention to how climate change affects historically marginalized communities and underrepresented audiences by continuing to provide pathways for including them as part of the development process of the NCA—making sure that they are heard throughout the development process—and more broadly by sustaining efforts to provide information about the climate-associated needs of those groups. The evaluators engaged by the Program should include persons whose backgrounds and lived experiences afford them understanding of underserved communities, including those exposed to the impacts of climate change.

Recommendation 5-4: Guided by its logic model, the U.S. Global Change Research Program (USGCRP) should progressively develop a roadmap of the network of networks in which the NCA and its related products are used. This roadmap would show the nodes that are hypothesized to play key roles in the diffusion of usable knowledge in the NCA and related products. The results should be considered when selecting which audiences to target in the evaluation.

A difficult issue is identifying and examining the underserved, particularly those who make no use of the NCA. If such audiences are involved in transmitting climate-related information, then a media search might identify them (depending on which media are included). If the audiences are users of climate-related information but not disseminators, then they may be hard to find (e.g., the NCA does not monitor who accesses its information, although such monitoring could be established). If they are neither disseminators nor users, even if their decisions might be improved substantially by using reliable information about climate change, then they are especially hard to identify and to get access to. One might possibly be able to contact individuals within these audiences and gain useful information, even without attempting to construct a statistically representative sample. The goal is to gain insight into what information is useful, even to those not now using knowledge of climate change to make decisions, and into better ways to build trust in the reliable science provided in the NCA.

CHOOSING APPROPRIATE METHODOLOGIES

Only after the audiences are chosen can the appropriate methodologies be determined. Some audiences might be examined using survey questionnaires; this works best when the members of an audience can be identified readily (e.g., using a membership list) and the questions to be asked are generally close-ended in nature (e.g., yes/no questions or multiple choice). Sometimes strategies for survey research are available even without a membership list (e.g., there is no comprehensive list of K–12 educators, but one could first sample schools and then teachers within schools), although cost would also be a consideration. If a more extended response is desired, then qualitative tools such as in-depth interviews and focus groups can be useful. Sometimes, these qualitative tools are used first in exploratory research to establish a general understanding of the issues, after which surveys might be used. Analysis of secondary (preexisting) data can also be useful without contacting any members of the audience, such as in a citation analysis. Case studies can be especially valuable for providing deeper insights and for exploratory research, focusing on a particular situation in depth rather than taking a broader view. Case studies can potentially use a mixture of tools, including surveys, in-depth interviews, focus groups, and analysis of secondary data.

Chapter 6 provides additional information about these methodologies, and then looks at five different audiences and how the evaluation approach might be customized for each audience.

Recommendation 6-1: The U.S. Global Change Research Program should design its evaluation and data collection plan so that the methods used for priority audiences can answer the overall evaluation

questions identified in the logic model. **The methods and approach chosen should be tailored to the audience and the evaluation question being investigated.**

CONTINUOUS EVALUATION AND IMPROVEMENT

For several reasons, the committee suggests that USGCRP adopt a policy of continuous evaluation and improvement rather than performing a one-time study. First, USGCRP has expressed an interest in using evaluation results to inform improvement, and in making evaluation a regular feature of its work. Second, USGCRP is operating in a rapidly changing landscape of climate services, and evaluation results can help to guide that evolution. Third, some aspects of effective evaluation for such a dynamic and cyclical program will benefit from conducting multiple iterations, such as an exploratory study followed by a survey, or different surveys to widen the range of audiences included.

The ability to support continuous improvement depends on what data are collected. For example, simply finding out what percentage of an audience makes use of the NCA provides little direct information toward program improvement, although it may indicate where there are gaps in reaching audiences. If the evaluation looks further into the reasons an audience has not used the NCA (e.g., lack of knowledge of the NCA, difficulties in finding information), then such information may help in knowing where and how to direct program improvement efforts. Iterative evaluations, then, can help determine whether program changes have had a positive effect.

Multiple evaluation iterations would help in other ways as well. USGCRP may initially have too little knowledge about some audiences to conduct anything more than exploratory research, perhaps using a case study. Based on the knowledge gained from the exploratory research, later evaluations may be able to examine those audiences more systematically. And, possibly, if some audiences cannot be examined because of practical limitations, later evaluations may be able to include them in place of audiences well studied through previous evaluations. Also, for audiences such as NGOs, findings from a case study may lead to ways to better identify other NGOs from whom information might be obtained.

Recommendation 7-2: The U.S. Global Change Research Program should sequence evaluation into manageable components, allowing for iterative testing and learning about how to best pursue evaluation over time. Sequenced components may include conducting evaluability assessments, piloting focused on certain agencies or chapters of the National Climate Assessment, picking low-hanging fruit first, or developing case studies.

Particularly if there are multiple evaluations, the value of obtaining feedback is multiplied. Such feedback can help in understanding previous findings and in refining the methodology to be used in the future.

Recommendation 7-3: In communication about evaluation efforts, the U.S. Global Change Research Program (USGCRP) should aim for active two-way communication with users. Communication mechanisms may include ongoing feedback, interim findings, meetings to tailor the communication of evaluation findings to particular situations, and communication about how input was used that helps connect evaluation efforts with USGCRP's objectives.

MULTISTEP APPROACH TO EVALUATION

To ensure that an evaluation addresses priority needs and opportunities for the Program and the wide range of audiences and decision-makers who use the NCA, engagement in and buy-in to the evaluation by USGCRP leadership is critical to determine the scope of the evaluation, the team implementing the evaluation, and the budget.

Recommendation 3-1: The leadership of the U.S. Global Change Research Program should engage from the start in defining the evaluation scope and should ensure that the leadership perspective, as well as the necessary evaluation expertise, is incorporated throughout the design and implementation of the evaluation.

Recommendation 7-4: The U.S. Global Change Research Program should consider bringing in outside expertise and research capabilities—such as through contractors, consultants, grantees, or interagency agreements—to assist in designing and implementing the evaluation.

This report outlines a multistep approach that USGCRP can follow to develop an evaluation of the NCA:

- USGCRP engages an evaluator. The evaluator should be experienced in designing and conducting large-scale evaluations and proposing methods to support continuous improvement.
- Working with the evaluator, USGCRP develops a logic model, spelling out its objectives for the NCA (see Chapter 3) and identifying important participants and audiences (see Chapters 4 and 5) for the information in the NCA and associated products; spelling out causal assumptions about the use of NCA information; and specifying the causal assumptions to be investigated in an evaluation (see Chapter 3). The development of a logic model and evaluation questions requires substantial involvement by USGCRP leadership.
- Working with important participants, including its federal agency members, USGCRP shares its logic model, and modifies it and the evaluation questions in light of comments from its partners (see Chapter 3).
- USGCRP's evaluator proposes an evaluation design: a process that engages selected audiences to provide data and identifies existing sources of information that will collectively provide insights to answer the evaluation questions (see Chapter 6). The design is likely to include a portfolio of methods, aimed at mapping the networks through which climate information travels and identifying groups to be investigated using evaluation instruments aimed at answering more specific questions.
- USGCRP incorporates information collection for evaluation purposes in its process of developing the sixth NCA. The Program also develops a process for continuous improvement, supported by ongoing information gathering and evaluation (see Chapter 7).
- USGCRP's evaluator collects new data and analyzes existing data. As the evaluation progresses, the evaluator consults with USGCRP on modifications to the evaluation design in light of initial findings. The data collection is likely to involve multiple phases, sometimes starting with exploratory analysis, and sometimes possibly requiring a long-term approach that extends beyond the development of a single NCA report. The evaluator prepares a summary of information and the answers to the evaluation questions.
- USGCRP drafts a response to the evaluation and engages participants and audiences via a public process, seeking comment on the evaluation findings and the Program's draft response. A final response is provided to the Office of Science & Technology Policy and, if requested, to Congress.

Evaluating the NCA is complex because of the many audiences that the assessment can serve in informing decision-making. The guidance developed in this report will require substantial effort and resources from USGCRP, commensurate with its influence and the need for authoritative and reliable information as the nation navigates the impacts of a changing climate. Investing in the evaluation of these critical products may hold substantial benefits. The concepts discussed here can increase the understanding of how the NCA is used, provide information needed to make future assessments more useful, and aid USGCRP in prioritizing assessment-related activities. The committee commends USGCRP for recognizing the need for evaluation by requesting this report; the recommendations advanced seek to respond to that need.

References

AEA (American Evaluation Association). 2011. *Public statement on cultural competence in evaluation*. Fairhaven, MA: AEA.

———, n.d. *What is evaluation?* https://www.eval.org/About/What-is-Evaluation.

Albright, K., P. Shah, M. Santodomingo, and J. Scandlyn. 2020. Dissemination of information about climate change by state and local public health departments: United States, 2019–2020. *American Journal of Public Health* 110(8):1184–1190. https://doi.org/10.2105/ajph.2020.305723.

Aleta, A., and Y. Moreno. 2019. Multilayer networks in a nutshell. *Annual Review of Condensed Matter Physics* 10(1):45–62. https://doi.org/10.1146/annurev-conmatphys-031218-013259.

Allan, R. P., P. A. Arias, S. Berger, J. G. Canadell, C. Cassou, C. Chen, A. Cherchi, S. L. Connors, E. Coppola, and F. A. Cruz. 2023. Intergovernmental Panel on Climate Change (IPCC). Summary for policymakers. In *Climate change 2021: The physical science basis*. Contribution of Working Group I to the Sixth Assessment Report of the Intergovernmental Panel on Climate Change. Cambridge, UK: Cambridge University Press.

Almasy, T., and D. Thompson. 2013. *Media relations strategy, tracking and evaluation essentials for nonprofit organizations*. Atlanta: Healthcare Georgia Foundation. https://search.issuelab.org/resources/28572/28572.pdf.

AMEC (International Association for the Measurement and Evaluation of Communication). 2024. *Integrated evaluation framework*. https://amecorg.com/amecframework.

Anstead, N. 2021. Social media in politics. In *The international encyclopedia of digital communication and society*, edited by R. Mansell. Hoboken, NJ: Wiley.

Aral, S. 2016. The future of weak ties. *American Journal of Sociology* 121(6):1931–1939. https://doi.org/10.1086/686293.

Arnstein, S. R. 1969. A ladder of citizen participation. *Journal of the American Institute of Planners* 35(4):216–224.

Avery, C. W., A. R. Crimmins, S. Basile, and A. Lustig. 2023. Appendix 1: Assessment development process. In *Fifth national climate assessment*, edited by A. R. Crimmins, C. W. Avery, D. R. Easterling, K. E. Kunkel, B. C. Stewart, and T. K. Maycock. Washington, DC: U.S. Global Change Research Program.

Avery, C. W., D. R. Reidmiller, T. S. Carter, K. L. M. Lewis, and K. Reeves. 2018. Report development process. In *Impacts, risks, and adaptation in the United States: Fourth national climate assessment, Volume II*, edited by D. R. Reidmiller, C. W. Avery, D. R. Easterling, K. E. Kunkel, K. L. M. Lewis, T. K. Maycock, and B. C. Stewart. Washington, DC: U.S. Global Change Research Program.

Balbach, E. D. 1999. *Using case studies to do program evaluation*. Sacramento: California Department of Health Services.

Bamzai-Dodson, A., A. E. Cravens, A. Wade, and R. A. McPherson. 2021. Engaging with stakeholders to produce actionable science: A framework and guidance. *Weather, Climate, and Society* 13(4):1027–1041. https://doi.org/10.1175/wcas-d-21-0046.1.

Bazzaz Abkenar, S., M. Haghi Kashani, E. Mahdipour, and S. M. Jameii. 2021. Big data analytics meets social media: A systematic review of techniques, open issues, and future directions. *Telematics and Informatics* 57:e101517. https://doi.org/10.1016/j.tele.2020.101517.

Behl, M., S. Carley, R. Connolly, and F. Niepold. 2015. Teaching about the ocean and coasts with the National Climate Assessment. *Current: The Journal of Marine Education* 29(4):18.

Bener, A. B., B. Çağlayan, A. D. Henry, and P. Prałat. 2016. Empirical models of social learning in a large, evolving network. *PLoS ONE* 11(10):e0160307. https://doi.org/10.1371/journal.pone.0160307.

Besley, J. C., and A. Dudo. 2022. *Strategic science communication: A guide to setting the right objectives for more effective public engagement.* Baltimore, MD: Johns Hopkins University Press.

Bianconi, G. 2023. *Multilayer networks: Structure and function.* Oxford, UK: Oxford University Press.

Bidwell, D., T. Dietz, and D. Scavia. 2013. Fostering knowledge networks for climate adaptation. *Nature Climate Change* 3:610–611.

Björneborn, L., and P. Ingwersen. 2004. Toward a basic framework for webometrics. *Journal of the American Society for Information Science and Technology* 55(14):1216–1227. https://doi.org/10.1002/asi.20077.

Blamey, A., and M. Mackenzie. 2007. Theories of change and realistic evaluation. *Evaluation* 13(4):439–455. https://doi.org/10.1177/1356389007082129.

Bloom, A. 2021, December 9. With great care, climate change is in the classroom. *The Nature Conservancy.* https://www.nature.org/en-us/about-us/who-we-are/how-we-work/youth-engagement/climate-change-education-us.

Blunden, J., T. Boyer, and E. Bartow-Gillies. 2023. State of the climate in 2022. *Bulletin of the American Meteorological Society* 104(9):S1–S516. https://doi.org/10.1175/2023bamsstateoftheclimate.1.

Borgatti, S. P. 2024. *Analyzing social networks.* Los Angeles, CA: Sage Publications.

Boyack, K. W., and R. Klavans. 2010. Co-citation analysis, bibliographic coupling, and direct citation: Which citation approach represents the research front most accurately? *Journal of the American Society for Information Science and Technology* 61(12):2389–2404. https://doi.org/10.1002/asi.21419.

Brass, D. J. 2022. New developments in social network analysis. *Annual Review of Organizational Psychology and Organizational Behavior* 9(1):225–246. https://doi.org/10.1146/annurev-orgpsych-012420-090628.

Braunschweiger, D. 2022. Cross-scale collaboration for adaptation to climate change: A two-mode network analysis of bridging actors in Switzerland. *Regional Environmental Change* 22(110). https://doi.org/10.1007/s10113-022-01958-4.

Brimicombe, C. 2022. Is there a climate change reporting bias? A case study of English-language news articles, 2017–2022. *Geoscience Communication* 5(3):281–287. https://doi.org/10.5194/gc-5-281-2022.

Brin, S., and L. Page. 1998. The anatomy of a large-scale hypertextual Web search engine. *Computer Networks and ISDN Systems* 30(1):107–117. https://doi.org/10.1016/S0169-7552(98)00110-X.

Buizer, J. L., P. Fleming, S. L. Hays, K. Dow, C. B. Field, D. Gustafson, A. Luers, and R. H. Moss. 2013. *Report on preparing the nation for change: Building a sustained National Climate Assessment process.* National Climate Assessment and Development Advisory Committee. https://downloads.globalchange.gov/nca/NCADAC/NCADAC_Sustained_Assessment_Special_Report_Sept2013.pdf.

Burt, R. S., and N. Celotto. 1992. The network structure of management roles in a large matrix firm. *Evaluation and Program Planning* 15(3):303–326. https://doi.org/10.1016/0149-7189(92)90095-C.

Bykov, I. 2021. On special issue: The Politization of Social Problems in Mass Media. *Journal of Political Research* 5(4):3–6. https://doi.org/10.12737/2587-6295-2021-5-4-3-6.

Campbell, E., S. Patzer, L. Beall, A. Gallagher, and E. Maibach. 2020. Using social science in National Park Service climate communications: A case study in the National Capital Region. *Parks Stewardship Forum* 36(1). https://doi.org/10.5070/p536146377.

Carrington, P. J., J. Scott, and S. Wasserman (Eds.). 2005. *Models and methods in social network analysis.* Cambridge, UK: Cambridge University Press.

Cash, D. W., W. C. Clark, F. Alcock, N. M. Dickson, N. Eckley, D. H. Guston, J. Jäger, and R. B. Mitchell. 2003. Knowledge systems for sustainable development. *Proceedings of the National Academy of Sciences* 100(14):8086–8091. https://doi.org/10.1073/pnas.1231332100.

Castells, M. 1996. *The rise of the network society.* Malden, MA: Blackwell.

CDC (Centers for Disease Control and Prevention). 1999. *Framework for program evaluation in public health.* https://www.cdc.gov/mmwr/PDF/rr/rr4811.pdf.

———. 2011. *Introduction to program evaluation for public health programs: A self-study guide.* https://www.cdc.gov/evaluation/guide/CDCEvalManual.pdf.

———. 2018. *Program evaluation framework checklist for step 2.* https://www.cdc.gov/evaluation/steps/step2/index.htm.

———. 2022. *Preferred terms for select population groups & communities.* https://www.cdc.gov/healthcommunication/Preferred_Terms.html.

———. 2023a. *Program evaluation.* https://www.cdc.gov/evaluation/index.htm.

REFERENCES

———. 2023b. *A Framework for Program Evaluation.* https://www.cdc.gov/evaluation/framework/index.htm.

CDC, n.d. *Practical use of program evaluation among sexually transmitted disease (STD) programs.* https://www.cdc.gov/std/program/pupestd.htm

Chen, H.-T. 2005. *Practical program evaluation: Assessing and improving planning, implementation, and effectiveness.* Thousand Oaks, CA: Sage.

Chen, W., C. Castillo, and L. V. S. Lakshmanan. 2022. *Information and influence propagation in social networks.* Cham: Springer International.

Cho, R. 2023. Climate education in the U.S.: Where it stands, and why it matters. *State of the Planet: News from the Columbia Climate School.* https://news.climate.columbia.edu/2023/02/09/climate-education-in-the-u-s-where-it-stands-and-why-it-matters/.

Chouinard, J. A. 2013. The case for participatory evaluation in an era of accountability. *American Journal of Evaluation* 34(2):237–253. https://doi.org/10.1177/1098214013478142.

Clarke, B., F. Otto, R. Stuart-Smith, and L. Harrington. 2022. Extreme weather impacts of climate change: An attribution perspective. *Environmental Research: Climate* 1(1):e012001. https://doi.org/10.1088/2752-5295/ac6e7d.

Cloyd, E., S. C. Moser, E. Maibach, J. Maldonado, and T. Chen. 2016. Engagement in the Third U.S. National Climate Assessment: Commitment, capacity, and communication for impact. *Climatic Change* 135(1):39–54. https://doi.org/10.1007/s10584-015-1568-y.

College Board. 2021. *AP Environmental Science overview.* https://apcentral.collegeboard.org/media/pdf/ap-environmental-science-course-overview.pdf.

Colorado Trust. 2002. *Guidelines and best practices for culturally competent evaluations.* https://www.coloradotrust.org/wp-content/uploads/2015/03/GuidelinesBestPracticesCulturally04.pdf.

Corbera, E., L. Calvet-Mir, H. Hughes and M. Paterson. 2016. Patterns of authorship in the IPCC Working Group III report. *Nature Climate Change* 6(1):94–99.

Cousins, J. B., and E. Whitmore. 1998. Framing participatory evaluation. *New Directions for Evaluation* 1998(80):5–23. https://doi.org/10.1002/ev.1114.

Crano, W. D., M. B. Brewer, and A. Lac. 2014. *Principles and methods of social research.* New York: Routledge. https://doi.org/10.4324/9781315768311.

Cummings, C. L. 2017. Science communication. In *The SAGE encyclopedia of communication research methods*, edited by M. Allen. Thousand Oaks, CA: Sage.

Cunningham, R., C. Cvitanovic, T. Measham, B. Jacobs, A.-M. Dowd, and B. Harman. 2016. Engaging communities in climate adaptation: The potential of social networks. *Climate Policy* 16(7):894–908. https://doi.org/10.1080/14693062.2015.1052955.

Cvitanovic, C., R. Cunningham, A. M. Dowd, S. M. Howden, and E. I. Van Putten. 2017. Using social network analysis to monitor and assess the effectiveness of knowledge brokers at connecting scientists and decision-makers: An Australian case study. *Environmental Policy and Governance* 27(3):256–269. https://doi.org/10.1002/eet.1752.

Dantzker, H., M. Shah, R. Gupta, C. Merse, J, Yoo, and K. Ward. 2016. *Evaluation of the Third NCA Production and Dissemination Processes and Products: Briefing to the Advisory Committee for the Sustained National Climate Assessment.* Presented at Bethesda, MD.

Declet-Barreto, J. 2024. *Union of Concerned Scientists' perspective on the National Climate Assessment.* Presentation at the 5th National Climate Assessment Information Gathering Session on February 28, 2024, Washington, DC.

Ding, Y., G. Zhang, T. Chambers, M. Song, X. Wang, and C. Zhai. 2014. Content-based citation analysis: The next generation of citation analysis. *Journal of the Association for Information Science and Technology* 65(9):1820–1833. https://doi.org/10.1002/asi.23256.

Djenontin, I. N. S., and A. M. Meadow. 2018. The art of co-production of knowledge in environmental sciences and management: Lessons from international practice. *Environmental Management* 61(6):885–903.

Dolšak, N., and A. Prakash. 2022. Three faces of climate justice. *Annual Review of Political Science* 25(1):283–301. https://doi.org/10.1146/annurev-polisci-051120-125514.

Dubois, B., and M. E. Krasny. 2016. Educating with resilience in mind: Addressing climate change in post-Sandy New York City. *Journal of Environmental Education* 47(4):255–270. https://doi.org/10.1080/00958964.2016.1167004.

Dudek, J., D. G. Pina, and R. Costas. 2021. Co-link analysis as a monitoring tool: A webometric use case to map the web relationships of research projects. *Proceedings of the 18th International Conference on Scientometrics & Informetrics*, pp. 339–344. https://doi.org/10.48550/arXiv.2110.04251.

EOP (Executive Office of the President. 2021. *Executive Order 14008 of January 27, 2021.* https://www.federalregister.gov/documents/2021/02/01/2021-02177/tackling-the-climate-crisis-at-home-and-abroad.

———. 2023. *A federal framework and action plan for climate services.* https://www.whitehouse.gov/wp-content/uploads/2023/03/ftac_report_03222023_508.pdf

Errett, N. A., K. Dolan, C. Hartwell, J. Vickery, and J. J. Hess. 2022. Adapting by their bootstraps: State and territorial public health agencies struggle to meet the mounting challenge of climate change. *American Journal of Public Health* 112(10):1379–1381. https://doi.org/10.2105/ajph.2022.307038.

EvalCommunity. 2024. *Evaluation glossary.* https://www.evalcommunity.com/evaluation-glossary/.

———, n.d. *Types of evaluation: Theory, case studies and job interview preparation.* https://www.evalcommunity.com/career-center/types-of-evaluation/.

Feiock, R. C., I. W. Lee, and H. J. Park. 2012. Administrators' and elected officials' collaboration networks: Selecting partners to reduce risk in economic development. *Public Administration Review* 72(Suppl.):S58–S68. https://doi.org/10.1111/j.1540-6210.2012.02659.x.

Fisher, D. R. 2012. Youth political participation: Bridging activism and electoral politics. *Annual Review of Sociology* 38(1):119–137. https://doi.org/10.1146/annurev-soc-071811-145439.

Fitzgerald, H. E. 2018. Community. In *The SAGE encyclopedia of lifespan human development*, edited by M. Bornstein and M. H. Bornstein. Los Angeles, CA: Sage.

Frank, K. A., and R. Xu. 2020. *Causal inference for social network analysis.* Oxford: Oxford University Press. https://doi.org/10.1093/oxfordhb/9780190251765.013.21.

Frank, K. A., T. Chen, E. Brown, A. Larsen, and W. B. J. Baule. 2023. A network intervention for natural resource management in the context of climate change. *Social Networks* 75:55–64. https://doi.org/10.1016/j.socnet.2022.03.003.

Frumkin, H., J. Hess, G. Luber, J. Malilay, and M. McGeehin. 2008. Climate change: The public health response. *American Journal of Public Health* 98(3):435–445. https://doi.org/10.2105/ajph.2007.119362.

GAO (General Accounting Office). 1987. *Case study evaluations.* https://www.gao.gov/products/132683.

GAO (Government Accountability Office) 2018. *Climate change: Analysis of reported federal funding.* https://www.gao.gov/assets/d18223.pdf.

Goodrich, K. A., K. D. Sjostrom, C. Vaughan, L. Nichols, A. Bednarek, and M. C. Lemos. 2020. Who are boundary spanners and how can we support them in making knowledge more actionable in sustainability fields? *Current Opinion in Environmental Sustainability* 42:45–51. https://doi.org/10.1016/j.cosust.2020.01.001.

Gorman, B. 2011. The death and life of American journalism: The media revolution that will begin the world again. *Canadian Journal of Communication* 35(4):634–636. https://doi.org/10.22230/cjc.2010v35n4a2414.

Granovetter, M. S. 1973. The strength of weak ties. *American Journal of Sociology* 78(6):1360–1380. http://www.jstor.org/stable/2776392.

Greene, J. C. 1988. Communication of results and utilization in participatory program evaluation. *Evaluation and Program Planning* 11(4):341–351. https://doi.org/10.1016/0149-7189(88)90047-X.

GSA and OMB (U.S. General Services Administration and Office of Management and Budget). n.d. *About the PRA.* https://pra.digital.gov/about.

Hall, J. A., M. Blair, J. L. Buizer, D. I. Gustafson, B. Holland, S. C. Moser, and A. M. Waple. 2014. Sustained assessment: A new vision for future U.S. assessments. In *Climate change impacts in the United States: The Third national Climate Assessment.* U.S. Global Change Research Program, pp. 719–726. http://nca2014.globalchange.gov/report/response-strategies/sustained-assessment.

Henry, A. D. 2016. *Network segregation and policy learning.* Oxford: Oxford University Press. https://doi.org/10.1093/oxfordhb/9780190228217.013.23.

———. 2020. Meeting the challenge of learning for sustainability through policy networks. *Human Ecology Review* 26(2):171–193.

———. 2023. Evaluating collaborative institutions by segregation and homophily in policy networks. *Public Administration* 101(2):604–621. https://doi.org/10.1111/padm.12800.

Henry, A. D., and B. Vollan. 2014. Networks and the challenge of sustainable development. *Annual Review of Environment and Resources* 39:583–610. https://doi.org/10.1146/annurev-environ-101813-013246.

Henry, A. D., K. Ingold, D. Nohrstedt, and C. M. Weible. 2014. Policy change in comparative contexts: Applying the advocacy coalition framework outside of western Europe and North America. *Journal of Comparative Policy Analysis: Research and Practice* 16(4):299–312. https://doi.org/10.1080/13876988.2014.941200.

Henry, A. D., T. Dietz, and R. L. Sweeney. 2021. Coevolution of networks and beliefs in U.S. environmental risk policy. *Policy Studies Journal* 49(3):675–702. https://doi.org/10.1111/psj.12407.

Hoffman, M., M. Lubell, and V. Hillis. 2014. Linking knowledge and action through mental models of sustainable agriculture. *Proceedings of the National Academy of Sciences* 111(36):13016–13021. https://doi.org/10.1073/pnas.1400435111.

REFERENCES

Hunter, L. Y. 2023. Social media, disinformation, and democracy: How different types of social media usage affect democracy cross-nationally. *Democratization* 30(6):1040–1072. https://doi.org/10.1080/13510347.2023.2208355.

IPCC (Intergovernmental Panel on Climate Change). 2021. *Climate change 2021—the physical science basis*. Working Group I contribution to the Sixth Assessment Report of the Intergovernmental Panel on Climate Change. Cambridge: Cambridge University Press. https://www.cambridge.org/core/product/415F29233B8BD19FB55F65E3DC67272B.

———. 2023. *Climate change 2023: Synthesis report*. Contribution of Working Groups I, II and III to the Sixth Assessment Report of the Intergovernmental Panel on Climate Change, edited by H. Lee and J. Romero. IPCC, Geneva, Switzerland, pp. 35–115. https://doi.org/10.59327/IPCC/AR6-9789291691647

Jacobs, K., S. Moser, and J. Buizer. 2016. *The U.S. National Climate Assessment*. Cham: Springer Climate. https://doi.org/10.1007/978-3-319-41802-5.

Jagannathan, K., J. C. Arnott, C. Wyborn, N. Klenk, K. J. Mach, R. H. Moss, and K. D. Sjostrom. 2020. Great expectations? Reconciling the aspiration, outcome, and possibility of co-production. *Current Opinion in Environmental Sustainability* 42:22–29. https://doi.org/10.1016/j.cosust.2019.11.010.

Jones, N. A., H. Ross, T. Lynam, P. Perez, and A. Leitch. 2011. Mental models: An interdisciplinary synthesis of theory and methods. *Ecology and Society* 16(1):46.

Joseph, H. A., E. Mallen, M. McLaughlin, E. Grossman, T. J. Holmes, A. Locklear, E. Powell, L. Thie, C. K. Uejio, K. Vacca, C. Williams, T. Bishop, C. Jeffers, H. Siegel, and C. Austin. 2023. Evaluating public health strategies for climate adaptation: Challenges and opportunities from the Climate-Ready States and Cities Initiative. *PLoS Climate* 2(3):e0000102. https://doi.org/10.1371/journal.pclm.0000102.

Julian, D. A. 1997. The utilization of the logic model as a system level planning and evaluation device. *Evaluation and Program Planning* 20(3):251–257. https://doi.org/10.1016/S0149-7189(97)00002-5.

Junge, K., J. Cullen, and G. Iacopini. 2020. Using contribution analysis to evaluate large-scale, transformation change processes. *Evaluation* 26(2):227–245. https://doi.org/10.1177/1356389020912270.

Kahneman, D. 2011. *Thinking, fast and slow*. New York: Farrar, Straus and Giroux.

Kalafatis, S. E., M. C. Lemos, Y.-J. Lo, and K. A. Frank. 2015. Increasing information usability for climate adaptation: The role of knowledge networks and communities of practice. *Global Environmental Change* 32:30–39.

Kallemeyn, L. M. 2009. Methodological changes and respecting stakeholder dignity. *American Journal of Evaluation* 30(4):575–580.

Kane, R., C. Levine, C. Orians and C. Reinelt. 2021. Contribution analysis: A promising method for assessing advocacy's impact. *New Directions for Evaluation* 2021:45–57. https://doi.org/10.1002/ev.20471.

Karl, T. R., J. M. P. Melillo, C. Thomas, and S. J. Hassol. 2009. *Global climate change impacts in the United States*. Cambridge, UK: Cambridge University Press.

Keith, R. E., J. C. Crosson, A. S. O'Malley, D. Cromp, and E. F. Taylor. 2017. Using the Consolidated Framework for Implementation Research (CFIR) to produce actionable findings: A rapid-cycle evaluation approach to improving implementation. *Implementation Science* 12(1):15. https://doi.org/10.1186/s13012-017-0550-7.

Kim, M., and R. M. Fernandez. 2023. What makes weak ties strong? *Annual Review of Sociology* 49(1):177–193. https://doi.org/10.1146/annurev-soc-030921-034152.

King, D., M. Griffin, and E. Bell. 2023. Inclusion and exclusion in management education and learning: A deliberative approach to conferences. *Academy of Management Learning & Education* 22(1):40–62. https://doi.org/10.5465/amle.2020.0089.

Kirchhoff, C. J., R. Esselman, and D. Brown. 2015. Boundary organizations to boundary chains: Prospects for advancing climate science application. *Climate Risk Management* 9:20–29. https://doi.org/10.1016/j.crm.2015.04.001.

Kirk, K. B., A. U. Gold, T. S. Ledley, S. B. Sullivan, C. A. Manduca, D. W. Mogk, and K. Wiese. 2014. Undergraduate climate education: Motivations, strategies, successes, and support. *Journal of Geoscience Education* 62(4):538–549.

Kivelä, M., A. Arenas, M. Barthelemy, J. P. Gleeson, Y. Moreno, and M. A. Porter. 2014. Multilayer networks. *Journal of Complex Networks* 2(3):203–271. https://doi.org/10.1093/comnet/cnu016.

Klaiman, T., A. Pantazis, A. Chainani, and B. Bekemeier. 2016. Using a positive deviance framework to identify local health departments in communities with exceptional maternal and child health outcomes: A cross-sectional study. *BMC Public Health* 16:602. https://doi.org/10.1186/s12889-016-3259-7.

Kong, X., Y. Shi, S. Yu, J. Liu, and F. Xia. 2019. Academic social networks: Modeling, analysis, mining and applications. *Journal of Network and Computer Applications* 132:86–103. https://doi.org/10.1016/j.jnca.2019.01.029.

Kreslake, J. M., M. Sarfaty, C. Roser-Renouf, A. A. Leiserowitz, and E. W. Maibach. 2018. The critical roles of health professionals in climate change prevention and preparedness. *American Journal of Public Health* 108(S2):S68–S69. https://doi.org/10.2105/ajph.2017.304044.

LaFrance, J., R. Nichols, and K. E. Kirkhart. 2012. Culture writes the script: On the centrality of context in indigenous evaluation. *New Directions for Evaluation* 2012(135):59–74.

Lam, S., and K. Skinner. 2021. The use of evaluability assessments in improving future evaluations: A scoping review of 10 years of literature (2008–2018). *American Journal of Evaluation* 42(4):523–540. https://doi.org/10.1177/1098214020936769.

Lance, P., and A. Hattori. 2016. *Sampling and evaluation: A guide to sampling for program impact evaluation*. Chapel Hill: MEASURE Evaluation, University of North Carolina.

Lee, I.-W., R. C. Feiock, and Y. Lee. 2012. Competitors and cooperators: A micro-level analysis of regional economic development collaboration networks. *Public Administration Review* 72(2):253–262. https://doi.org/10.1111/j.1540-6210.2011.02501.x.

Leiserowitz, A., E. Maibach, S. Rosenthal, J. Kotcher, E. Goddard, J. Carman, M. Verner, M. Ballew, J. Marlon, S. Lee, T. Myers, M. Golding, N. Badullovich, and K. Thier. 2023. *Global warming's six Americas, Fall 2023*. Yale Program on Climate Change Communication. https://climatecommunication.yale.edu/publications/global-warmings-six-americas-fall-2023/.

Lemos, M. C., C. J. Kirchhoff, S. E. Kalafatis, D. Scavia, and R. B. Rood. 2014. Moving climate information off the shelf: boundary chains and the role of RISAs as adaptive organizations. *Weather, Climate, and Society* 6(2):273–285. https://doi.org/10.1175/wcas-d-13-00044.1.

Lewin, K. 1946. Action research and minority problems. *Journal of Social Issues* 2(4):34–46.

Lopez Hernandez, A., J. L. Weinberg, A. El-Harakeh, L. Adeyemi, N. Potharaj, N. Oomman, and A. Kalbarczyk. 2022. Connectedness is critical: A social network analysis to support emerging women leaders in global health. *Annals of Global Health* 88(1):64. https://doi.org/10.5334/aogh.3811.

Marino, E. K., K. Maxwell, E. Eisenhauer, A. Zycherman, C. Callison, E. Fussell, M. D. Hendricks, F. H. Jacobs, A. Jerolleman, A. K. Jorgenson, E. M. Markowitz, S. T. S. Marquart-Pyatt, Melissa, R. L. Shwom, and K. Whyte. 2023. Social systems and justice. In *Fifth National Climate Assessment*, edited by A. R. Crimmins, C. W. Avery, D. R. Easterling, K. E. Kunkel, B. C. Stewart, and T. K. Maycock. Washington, DC: U.S. Global Change Research Program.

Marinucci, G., G. Luber, C. Uejio, S. Saha, and J. Hess. 2014. Building resilience against climate effects: A novel framework to facilitate climate readiness in public health agencies. *International Journal of Environmental Research and Public Health* 11(6):6433–6458. https://doi.org/10.3390/ijerph110606433.

Masuda, Y. J., Y. Liu, S. M. W. Reddy, K. A. Frank, K. Burford, J. R. B. Fisher, and J. Montambault. 2018. Innovation diffusion within large environmental NGOs through informal network agents. *Nature Sustainability* 1(4):190–197. https://doi.org/10.1038/s41893-018-0045-9.

Mayne, J. 2008. Contribution analysis: An approach to exploring cause and effect. ILAC Brief 16:e52525. Institutional Learning and Change Initiative, Biodiversity International.

Mayne, J. 2011. Contribution analysis: Addressing cause and effect. In *Evaluating the Complex*, K. Forss, M. Marra and R. Schwartz (Eds.), pp. 53–96. Transaction.

Mayne, J. 2012. Contribution analysis: Coming of age? *Evaluation* 18(3):270–280. https://doi.org/10.1177/1356389012451663.

McCaston, M. K. 2005. *Tips for collecting, reviewing, and analyzing secondary data*. NY Health Foundation. https://nyhealthfoundation.org/wp-content/uploads/2019/02/Tips_for_Collecting_Reviewing_and_Analyz.pdf.

McLevey, J., J. Scott, and P. J. Carrington. 2023. *The Sage handbook of social network analysis*. Los Angeles: Sage.

McNamara, K. 2008. *Key concepts in urban geography*, edited by A. Latham, D. McCormack, K. McNamara, and D. McNeill. Thousand Oaks, CA: Sage.

Melillo, J. M., G. W. Yohe, and T. Richmond. 2014. *Climate change impacts in the United States: The Third National Climate Assessment*. U.S. Global Change Research Program. https://www.nrc.gov/docs/ML1412/ML14129A233.pdf.

Meyer, R. 2011. The public values failures of climate science in the US. *Minerva* 49(1):47-70. https://doi.org/10.1007/s11024-011-9164-4.

Mitchell, R. 2018. *Web scraping with Python: Collecting more data from the modern web*. Sebastopol, CA: O'Reilly Media.

MECCE (Monitoring and Evaluating Climate Communication and Education Project). 2022. *Mapping the landscape of K–12 climate change education policy in the United States*. https://dg56ycbvljkqr.cloudfront.net/sites/default/files/eepro-post-files/K-12%20Change.Full%20Report.pdf.

Monroe, M. C., R. R. Plate, A. Oxarart, A. Bowers, and W. A. Chaves. 2019. Identifying effective climate change education strategies: A systematic review of the research. *Environmental Education Research* 25(6):791–812. https://doi.org/10.1080/13504622.2017.1360842.

Montague, S. 2012. Theory-based approaches for practical evaluation. Presented at Canadian Evaluation Society Annual Learning Event, February 21, 2012.

Montgomery, B. L. 2018. Building and sustaining diverse functioning networks using social media and digital platforms to improve diversity and inclusivity. *Frontiers in Digital Humanities* 5. https://doi.org/10.3389/fdigh.2018.00022.

Morgan, M. G., R. Cantor, W. C. Clark, A. Fisher, H. D. Jacoby, A. C. Janetos, A. P. Kinzig, J. Melillo, R. B. Street, and T. J. Wilbanks. 2005. Learning from the U.S. national assessment of climate change impacts. *Environmental Science & Technology* 39(23):9023–9032. https://doi.org/10.1021/es050865i.

Morton, S. 2015. Creating research impact: The roles of research users in interactive research mobilization. *Evidence & Policy* 11(1):35–55. https://doi.org/10.1332/174426514x13976529631798.

Morton, S., and A. Cook. 2023. *How do you know if you are making a difference? A practical handbook for public service organisations.* Bristol, UK: Policy Press.

Moser, S. C. 2005. *Stakeholder involvement in the first US national assessment of the potential consequences of climate variability and change: An evaluation, finally.* Research report prepared for National Research Council, Committee on Human Dimensions of Global Change, Public Participation in Environmental Assessment and Decision Making, NAS/NRC, Washington, DC. http://www.susannemoser.com/documents/Moser_Draft_2-6-05.pdf.

Moss, R. H., S. Avery, K. Baja, M. Burkett, A. M. Chischilly, J. Dell, P. A. Fleming, K. Geil, K. Jacobs, A. Jones, K. Knowlton, J. Koh, M. C. Lemos, J. Melillo, R. Pandya, T. C. Richmond, L. Scarlett, J. Snyder, M. Stults, A. M. Waple, J. Whitehead, D. Zarrilli, B. M. Ayyub, J. Fox, A. Ganguly, L. Joppa, S. Julius, P. Kirshen, R. Kreutter, A. McGovern, R. Meyer, J. Neumann, W. Solecki, J. Smith, P. Tissot, G. Yohe, and R. Zimmerman. 2019. Evaluating knowledge to support climate action: A framework for sustained assessment: Report of an Independent Advisory Committee on Applied Climate Assessment. *Weather, Climate, and Society* 11(3):465–487. https://doi.org/10.1175/wcas-d-18-0134.1.

NASEM (National Academies of Science, Engineering, and Medicine). 2016. *Attribution of extreme weather events in the context of climate change.* Washington, DC: The National Academies Press.

———. 2017. *Accomplishments of the U.S. Global Change Research Program.* Washington, DC: The National Academies Press. https://doi.org/10.17226/24670.

———. 2023. *Review of the draft: Fifth National Climate Assessment.* Washington, DC: The National Academies Press.

NAST (National Assessment Synthesis Team). 2000. *Climate change impacts on the United States: The potential consequences of climate variability and change.* Cambridge, UK: Cambridge University Press.

NCEI (National Centers for Environmental Information). 2024. U.S. billion-dollar weather and climate disasters (2024). https://www.ncei.noaa.gov/access/billions/.

NCVO (National Council for Voluntary Organizations). 2023. *Using secondary data.* https://www.ncvo.org.uk/help-and-guidance/strategy-and-impact/impact-evaluation/planning-your-impact-and-evaluation/choosing-evaluation-methods/using-secondary-data.

Neuman, A., N. Shahor, I. Shina, A. Sarid, and Z. Saar. 2013. Evaluation utilization research: Developing a theory and putting it to use. *Evaluation and Program Planning* 36(1):64–70. https://doi.org/10.1016/j.evalprogplan.2012.06.001.

Newman, M. 2018. *Networks.* Oxford, UK: Oxford University Press.

Nisbet, M. C. 2009. Communicating climate change: Why frames matter for public engagement. *Environment: Science and Policy for Sustainable Development* 51(2):12–23. https://doi.org/10.3200/ENVT.51.2.12-23.

NOAA (National Oceanic and Atmospheric Association). 2024b. *Coordinated federal climate networks to enhance adaptation and resilience at the regional scale.* https://toolkit.climate.gov/content/federal-agency-coordination.

———. 2024c. *Regional climate centers.* https://www.ncei.noaa.gov/regional/regional-climate-centers.

NRC (National Research Council). 2007. *Analysis of global change assessments: Lessons learned.* Washington, DC: The National Academies Press.

———. 2008. *Public participation in environmental assessment and decision making.* Washington, DC: The National Academies Press. https://doi.org/10.17226/12434.

———. 2013. *A review of the draft: 2013 National Climate Assessment.* Washington, DC: The National Academies Press. https://doi.org/10.17226/18322.

NSTC (National Science and Technology Council). 2023. *A federal framework and action plan for climate services.* https://www.whitehouse.gov/wpcontent/uploads/2023/03/FTAC_Report_03222023_508.pdf.

O'Reilly, J. L., M. Vardy, K. D. Pryck, and M. D. S. F. Benedetti. 2024. *Inside the IPCC: How assessment practices shape climate knowledge.* Cambridge, UK: Cambridge University Press. https://www.cambridge.org/core/product/1699F8274D52678F74FCAA3B65B13E1B.

Oliver, S., L. Clarke-Jones, R. Rees, R. Milne, P. Buchanan, J. Gabbay, G. Gyte, A. Oakley, and K. Stein. 2004. Involving consumers in research and development agenda setting for the NHS: Developing an evidence-based approach. *Health Technology Assessment* 8(15). https://doi.org/10.3310/hta8150.

Oliver, S. R., R. W. Rees, L. Clarke-Jones, R. Milne, A. R. Oakley, J. Gabbay, K. Stein, P. Buchanan, and G. Gyte. 2008. A multidimensional conceptual framework for analyzing public involvement in health services research. *Health Expectations* 11(1):72–84.

OMB (Office of Management and Budget). 2022. *President Biden's FY 2023 budget reduces energy costs, combats the climate crisis, and advances environmental justice.* https://www.whitehouse.gov/omb/briefing-room/2022/03/28/president-bidens-fy-2023-budget-reduces-energy-costs-combats-the-climate-crisis-and-advances-environmental-justice/.

———. n.d. *Federal collection of information.* https://www.whitehouse.gov/omb/information-regulatory-affairs/federal-collection-information/#IIPR.

Orlove, B., P. Sherpa, N. Dawson, I. Adelekan, W. Alangui, R. Carmona, D. Coen, M. K. Nelson, V. Reyes-García, and J. Rubis. 2023. Placing diverse knowledge systems at the core of transformative climate research. *Ambio* 52(9):1431–1447.

OSTP and FEMA (Office of Science and Technology Policy and the Federal Emergency Management Agency). 2021. *Opportunities for expanding and improving climate information and services for the public: A report to the National Climate Task Force.* https://downloads.globalchange.gov/reports/eo-14008-211-d-report.pdf.

Palinkas, L. A., S. M. Horwitz, C. A. Green, J. P. Wisdom, N. Duan, and K. Hoagwood. 2015. Purposeful sampling for qualitative data collection and analysis in mixed method implementation research. *Administration and Policy in Mental Health and Mental Health Services Research* 42(5):533–544. https://doi.org/10.1007/s10488-013-0528-y.

Palomino, F., F. Paz, and A. Moquillaza. 2021. Web analytics for user experience: A systematic literature review. In *Design, User Experience, and Usability: UX Research and Design*, M. M. Soares, E. Rosenzweig, and A. Marcus (Eds.) Cham: Springer, pp. 312–326. https://doi.org/10.1007/978-3-030-78221-4_21.

Paradis, E., B. O'Brien, L. Nimmon, G. Bandiera, and M. A. Martimianakis. 2016. Design: Selection of data collection methods. *Journal of Graduate Medical Education* 8(2):263–264. https://doi.org/10.4300/jgme-d-16-00098.1.

Parson, E. A., R. W. Corell, E. J. Barron, V. Burkett, A. Janetos, L. Joyce, T. R. Karl, M. C. Maccracken, J. Melillo, M. G. Morgan, D. S. Schimel, and T. Wilbanks. 2003. Understanding climatic impacts, vulnerabilities, and adaptation in the United States: Building a capacity for assessment. *Climatic Change* 57(1/2):9–42. https://doi.org/10.1023/a:1022188519982.

Pasquini, G., and B. Kennedy. 2023. *Americans continue to have doubts about climate scientists' understanding of climate change.* Pew Research Center. https://www.pewresearch.org/short-reads/2023/10/25/americans-continue-to-have-doubts-about-climate-scientists-understanding-of-climate-change/.

Pasquini, G., A. Spencer, A. Tyson, and C. Funk. 2023. *Why some Americans do not see urgency on climate change.* Pew Research Center. https://www.pewresearch.org/science/2023/08/09/why-some-americans-do-not-see-urgency-on-climate-change/.

Patton, M. Q. (2003). Utilization-focused evaluation. In: *International handbook of educational evaluation.* Dordrecht: Springer. T. Kellaghan, and D. L. Stufflebeam (Eds.). https://doi.org/10.1007/978-94-010-0309-4_15.

———. 2012. *Essentials of utilization-focused evaluation.* Sage.

———. 2013. Utilization-focused evaluation (U-FE) Checklist. Western Michigan University. https://wmich.edu/sites/default/files/attachments/u350/2014/UFE_checklist_2013.pdf.

Perlino, C. 2006. *The public health workforce shortage: Left unchecked, will we be protected?* American Public Health Association. https://www.apha.org/media/files/pdf/factsheets/publichealthworkforceissuebrief.ashx.

Petzold, J., T. Hawxwell, K. Jantke, E. Gonçalves Gresse, C. Mirbach, I. Ajibade, S. Bhadwal, K. Bowen, A. P. Fischer, E. T. Joe, C. J. Kirchhoff, K. J. Mach, D. Reckien, A. C. Segnon, C. Singh, N. Ulibarri, D. Campbell, E. Cremin, L. Färber, G. Hegde, J. Jeong, A. M. Nunbogu, H. K. Pradhan, L. S. Schröder, M. A. R. Shah, P. Reese, F. Sultana, C. Tello, J. Xu, M. Garschagen, and The Global Adaptation Mapping Initiative. 2023. A global assessment of actors and their roles in climate change adaptation. *Nature Climate Change* 13(11):1250–1257. https://doi.org/10.1038/s41558-023-01824-z.

Pew Research Center. 2020. *Two-thirds of Americans think government should do more on climate.* https://www.pewresearch.org/science/wpcontent/uploads/sites/16/2020/06/PS_2020.06.23_government-climate_REPORT.pdf.

Popelier, L. 2018. A scoping review on the current and potential use of social network analysis for evaluation purposes. *Evaluation* 24(3):325–352. https://doi.org/10.1177/1356389018782219.

Puttick, S., and I. Talks. 2021. Teachers' sources of information about climate change: A scoping review. *The Curriculum Journal* 33(3):378–395. https://doi.org/10.1002/curj.136.

Rajkumar, K., G. Saint-Jacques, I. Bojinov, E. Brynjolfsson, and S. Aral. 2022. A causal test of the strength of weak ties. *Science* 377(6612):1304–1310. https://doi.org/10.1126/science.abl4476.

Reed, M. S., B. G. Merkle, E. J. Cook, C. Hafferty, A. P. Hejnowicz, R. Holliman, I. D. Marder, U. Pool, C. M. Raymond, K. E. Wallen, D. Whyte, M. Ballesteros, S. Bhanbhro, S. Borota, M. L. Brennan, E. Carmen, E. A. Conway, R. Everett, F. Armstrong-Gibbs, E. Jensen, G. Koren, J. Lockett, P. Obani, S. O'Connor, L. Prange, J. Mason, S. Robinson, P. Shukla, A. Tarrant, A. Marchetti, and M. Stroobant. 2024. Reimagining the language of engagement in a post-stakeholder world. *Sustainability Science* 19:1481–1490. https://doi.org/10.1007/s11625-024-01496-4.

Revelle, R., and H. E. Suess. 1957. Carbon dioxide exchange between atmosphere and ocean and the question of an increase of atmospheric CO^2 during the past decades. *Tellus* 9(1):18–27. https://doi.org/10.1111/j.2153-3490.1957.tb01849.x.

Robins, G., D. Lusher, C. Broccatelli, D. Bright, C. Gallagher, M. A. Karkavandi, P. Matous, J. Coutinho, P. Wang, and J. Koskinen. 2023. Multilevel network interventions: Goals, actions, and outcomes. *Social Networks* 72:108–120.

Roca-Barceló, A., M. B. Rice, Y. Nunez, G. Thurston, G. Weinmayr, K. Straif, C. Roscoe, K. L. Ebi, Z. J. Andersen, A. de Nazelle, and M. Negev. 2024. Climate action has valuable health benefits. *Environmental Epidemiology* 8(1):e288. https://doi.org/10.1097/ee9.0000000000000288.

Rodriguez-Franco, C., and T. J. Haan. 2015. Understanding climate change perceptions, attitudes, and needs of Forest Service resource managers. *Journal of Sustainable Forestry* 34(5):423–444.

Rogers, P. J. 2008. Using programme theory to evaluate complicated and complex aspects of interventions. *Evaluation* 14(1):29–48. https://doi.org/10.1177/1356389007084674.

Romanello, M., C. D. Napoli, C. Green, H. Kennard, P. Lampard, D. Scamman, M. Walawender, Z. Ali, N. Ameli, S. Ayeb-Karlsson, P. J. Beggs, K. Belesova, L. Berrang Ford, K. Bowen, W. Cai, M. Callaghan, D. Campbell-Lendrum, J. Chambers, T. J. Cross, K. R. van Daalen, C. Dalin, N. Dasandi, S. Dasgupta, M. Davies, P. Dominguez-Salas, R. Dubrow, K. L. Ebi, M. Eckelman, P. Ekins, C. Freyberg, O. Gasparyan, G. Gordon-Strachan, H. Graham, S. H. Gunther, I. Hamilton, Y. Hang, R. Hänninen, S. Hartinger, K. He, J. Heidecke, J. J. Hess, S. C. Hsu, L. Jamart, S. Jankin, O. Jay, I. Kelman, G. Kiesewetter, P. Kinney, D. Kniveton, R. Kouznetsov, F. Larosa, J. K. W. Lee, B. Lemke, Y. Liu, Z. Liu, M. Lott, M. Lotto Batista, R. Lowe, M. Odhiambo Sewe, J. Martinez-Urtaza, M. Maslin, L. McAllister, C. McMichael, Z. Mi, J. Milner, K. Minor, J. C. Minx, N. Mohajeri, N. C. Momen, M. Moradi-Lakeh, K. Morrissey, S. Munzert, K. A. Murray, T. Neville, M. Nilsson, N. Obradovich, M. B. O'Hare, C. Oliveira, T. Oreszczyn, M. Otto, F. Owfi, O. Pearman, F. Pega, A. Pershing, M. Rabbaniha, J. Rickman, E. J. Z. Robinson, J. Rocklöv, R. N. Salas, J. C. Semenza, J. D. Sherman, J. Shumake-Guillemot, G. Silbert, M. Sofiev, M. Springmann, J. D. Stowell, M. Tabatabaei, J. Taylor, R. Thompson, C. Tonne, M. Treskova, J. A. Trinanes, F. Wagner, L. Warnecke, H. Whitcombe, M. Winning, A. Wyns, M. Yglesias-González, S. Zhang, Y. Zhang, Q. Zhu, P. Gong, H. Montgomery, and A. Costello. 2023. The 2023 report of the Lancet: Countdown on health and climate change: The imperative for a health-centred response in a world facing irreversible harms. *Lancet* 402(10419):2346–2394. https://doi.org/10.1016/s0140-6736(23)01859-7.

Roozenbeek, J., S. Van Der Linden, and T. Nygren. 2020. Prebunking interventions based on the psychological theory of "inoculation" can reduce susceptibility to misinformation across cultures. *Harvard Kennedy School Misinformation Review*. https://doi.org/10.37016//mr-2020-008.

Rosenstock, L., C. Olenec, and G. R. Wagner. 1998. The National Occupational Research Agenda: A model of broad stakeholder input into priority setting. *American Journal of Public Health* 88(3):353–356. https://doi.org/10.2105/AJPH.88.3.353.

Rudolph, L., and S. Gould. 2015. Climate Change and health inequities: A framework for action. *Annals of Global Health* 81(3):432. https://doi.org/10.1016/j.aogh.2015.06.003.

Salganik, M. J. 2019. *Bit by bit: Social research in the digital age*. Princeton, NJ: Princeton University Press.

Schulte-Uebbing, L., G. Hansen, A. M. Hernández, and M. Winter. 2015. Chapter scientists in the IPCC AR5: Experience and lessons learned. *Current Opinion in Environmental Sustainability* 14:250–256. https://doi.org/10.1016/j.cosust.2015.06.012.

Sheehan, M. C., M. A. Fox, C. Kaye, and B. Resnick. 2017. Integrating health into local climate response: Lessons from the U.S. CDC Climate-Ready States and Cities Initiative. *Environmental Health Perspectives* 125(9):094501. https://doi.org/10.1289/ehp1838.

Sibbald, S. L., J. C. MacGregor, M. Surmacz, and C. N. Wathen. 2015. Into the gray: A modified approach to citation analysis to better understand research impact. *Journal of the Medical Library Association* 103(1):49–54. https://doi.org/10.3163/1536-5050.103.1.010.

Simons, H. 2021. Media and politics. In *Encyclopedia of American studies*. Baltimore, MD: Johns Hopkins University Press.

Smit, L. C., J. Dikken, M. J. Schuurmans, N. J. De Wit, and N. Bleijenberg. 2020. Value of social network analysis for developing and evaluating complex healthcare interventions: A scoping review. *BMJ Open* 10(11). https://doi.org/10.1136/bmjopen-2020-039681.

Smith, A. B. 2020. *U.S. billion-dollar weather and climate disasters, 1980 - present* (NCEI Accession 0209268).

Smith, P. A. 2001. Action learning and reflective practice in project environments that are related to leadership development. *Management Learning* 32(1):31–48.

Steg, L. 2023. Psychology of climate change. *Annual Review of Psychology* 74:391–421. https://doi.org/10.1146/annurev-psych-032720-042905.

Stern, A., S. Guckenburg, H. Persson, A. Petrosino, and J. Poirier. 2019. *Reflections on applying principles of equitable evaluation*. San Francisco, CA: WestEd.

Taherdoost, H. 2021. Data collection methods and tools for research: A step-by-step guide to choose data collection technique for academic and business research projects. *International Journal of Academic Research in Management* 10(1):10–38.

Tart, S., M. Groth, and P. Seipold. 2020. Market demand for climate services: An assessment of users' needs. *Climate Services* 17:100109. https://doi.org/10.1016/j.cliser.2019.100109.

Thelwall, M. 2012. A history of webometrics. *Bulletin of the American Society for Information Science and Technology* 38(6):18–23. https://doi.org/10.1002/bult.2012.1720380606.

Thomann, E., and M. Maggetti. 2020. Designing research with qualitative comparative analysis (QCA): Approaches, challenges, and tools. *Sociological Methods & Research* 49(2):356–386. https://doi.org/10.1177/0049124117729700.

Thomas, M. S. 2020. Equity in program evaluation: Equity as a measure in program evaluation. School of Professional and Continuing Studies Nonprofit Studies Capstone Projects. University of Richmond. https://scholarship.richmond.edu/spcs-nonprofitstudies-capstones/21/

Treen, K. M. d'I., H. T. P. Williams, and S. J. O'Neill. 2020. Online misinformation about climate change. *WIREs Climate Change* 11(5). https://doi.org/10.1002/wcc.665.

Trevisan, M. S. 2007. Evaluability assessment from 1986 to 2006. *American Journal of Evaluation* 28(3):290–303. https://doi.org/10.1177/1098214007304589.

Tripati, A., M. Shepherd, V. Morris, K. Andrade, K. P. Whyte, D. M. David-Chavez, J. Hosbey, J. E. Trujillo-Falcón, B. Hunter, D. Hence, D. Carlis, V. Brown, W. L. Parker, A. Geller, A. Reich, and M. Glackin. 2024. Centering equity in the nation's weather, water, and climate services. *Environmental Justice* 17(1):45–53. https://doi.org/10.1089/env.2022.0048.

Tsang, E. W. K. 2014. Generalizing from research findings: The merits of case studies. *International Journal of Management Reviews* 16(4):369–383. https://doi.org/10.1111/ijmr.12024.

Turek-Hankins, L. L., E. Coughlan de Perez, G. Scarpa, R. Ruiz-Diaz, P. N. Schwerdtle, E. T. Joe, E. K. Galappaththi, E. M. French, S. E. Austin, C. Singh, M. Siña, A. R. Siders, M. K. van Aalst, S. Templeman, A. M. Nunbogu, L. Berrang-Ford, T. Agrawal, the Global Systematic Review of Implemented Action Team, and K. J. Mach. 2021. Climate change adaptation to extreme heat: A global systematic review of implemented action. *Oxford Open Climate Change* 1(1). https://doi.org/10.1093/oxfclm/kgab005.

Turrentine, J. 2022. *Climate misinformation on social media is undermining climate action*. Natural Resources Defense Council. https://www.nrdc.org/stories/climate-misinformation-social-media-undermining-climate-action.

Tyson, A., and B. Kennedy. 2023. *How Americans view future harms from climate change in their community and around the U.S.* Pew Research Center. https://www.pewresearch.org/science/2023/10/25/how-americans-view-future-harms-from-climate-change-in-their-community-and-around-the-u-s/.

Tyson, A., C. Funk, and B. Kennedy. 2023. *What the data says about Americans' views of climate change*. Pew Research Center. https://www.pewresearch.org/short-reads/2023/08/09/what-the-data-says-about-americans-views-of-climate-change/.

University of Wisconsin Extension. 2024. *Enhancing program performance with logic models*. https://logicmodel.extension.wisc.edu/

U.S. Climate Resilience Toolkit. n.d. *Interagency collaboration: Three federal climate networks partner to enhance regional adaptation and resilience*. https://toolkit.climate.gov/sites/default/files/CPO_InteragencyCollaboration_20220105%20-%20FINAL.pdf.

USAID. 2013. Technical note: Evaluation case studies. https://usaidlearninglab.org/system/files/resource/files/usaid_case_study_tech_note_2013.pdf.

USDA (U.S. Department of Agriculture). 2021. *USDA climate hubs strategic plan 2020-2025*. https://www.climatehubs.usda.gov/sites/default/files/Strategic_Plan_Climate_Hubs_2020_2025_FINAL_v1.pdf

———. 2023a. *Climate hubs quarterly reporting*. https://www.climatehubs.usda.gov/climate-hubs-quarterly-reporting.

———. 2023b. *The USDA climate hubs: A national resource for climate science, tools, and information*. https://www.climatehubs.usda.gov/hubs/northwest/news/usda-climate-hubs-national-resource-climate-science-tools-and-information.

———. 2024. *Environmental justice*. https://www.climatehubs.usda.gov/priorities/environmental-justice?page=1.

———. n.d.-a. *Northwest adaptation in action*. https://www.climatehubs.usda.gov/hubs/northwest/topic/northwest-adaptation-action.

———. n.d.-b. *Northwest vulnerability assessments*. https://www.climatehubs.usda.gov/hubs/northwest/topic/northwest-vulnerability-assessments.

———. n.d.-c. *Welcome to the USDA Northwest Climate Hub*. https://www.climatehubs.usda.gov/hubs/northwest/tools.

USGCRP (U.S. Global Change Research Program). 2014. *Frameworks for evaluating the National Climate Assessment: Workshop report*. https://downloads.globalchange.gov/nca/workshop-reports/NCA-Evaluation-Frameworks-Workshop-Report-2014-final.pdf.

———. 2016a. *The NCA in review: Frameworks for evaluation*. https://www.globalchange.gov/highlights/nca-review-frameworks-evaluation.

———. 2016b. *The impacts of climate change on human health in the United States: A scientific assessment*. https://health2016.globalchange.gov.

———. 2018. *Summary findings: Impacts, risks, and adaptation in the United States: Fourth National Climate Assessment, Volume II*. https://nca2018.globalchange.gov.

———. 2023b. *Fifth National Climate Assessment public comment period & National Academies review annotation*. https://nca2023.globalchange.gov/downloads/NCA5_Public-Comments-Author-Responses-TOD.pdf.

———. 2023c. *Fifth national climate assessment*, edited by A. R. Crimmins, C. W. Avery, D. R. Easterling, K. E. Kunkel, B. C. Stewart, and T. K. Maycock. Washington, DC: U.S. Global Change Research Program. https://doi.org/10.7930/NCA5.2023.

———. 2023e. Front matter. In *Fifth National Climate Assessment*, Crimmins, A. R., C. W. Avery, D. R. Easterling, K. E. Kunkel, B. C. Stewart, and T. K. Maycock, Eds. U.S. Global Change Research Program, Washington, DC. https://doi.org/10.7930/NCA5.2023.FM.

———. 2024. *Our changing planet: The U.S. Global Change Research Program for fiscal year 2024*. https://downloads.globalchange.gov/ocp/ocp2024/Our-Changing-Planet_FY2024.pdf.

———. n.d. *Interagency crosscutting group on climate change and health*. https://globalchange.gov/our-work/interagency-groups/cchhg.

Valente, T. W. 2012. Network interventions. *Science* 337(6090):49–53. https://doi.org/10.1126/science.1217330.

Valente, T. W. 2017. Putting the network in network interventions. *Proceedings of the National Academy of Sciences* 114(36):9500-9501. https://doi.org/10.1073/pnas.1712473114.

Van Der Linden, S., A. Leiserowitz, S. Rosenthal, and E. Maibach. 2017. Inoculating the public against misinformation about climate change. *Global Challenges* 1(2):e1600008. https://doi.org/10.1002/gch2.201600008.

Vartanian, T. P. 2010. *Secondary data analysis*. Oxford, UK: Oxford University Press.

Vasileiadou, E., G. Heimeriks, and A. C. Petersen. 2011. Exploring the impact of the IPCC assessment reports on science. *Environmental Science & Policy* 14(8):1052–1061. https://doi.org/10.1016/j.envsci.2011.07.002.

Vasquez, K. S., S. Chatterjee, C. Khalida, D. Moftah, B. D'Orazio, A. Leinberger-Jabari, J. N. Tobin, and R. G. Kost. 2020. Using attendance data for social network analysis of a community-engaged research partnership. *Journal of Clinical and Translational Science* 5(1):e75. https://doi.org/10.1017/cts.2020.571.

Vedung, E. 2017. *Public policy and program evaluation*. New York: Routledge. https://doi.org/10.4324/9781315127767.

Venturini, T., K. D. Pryck, and R. Ackland. 2023. Bridging in network organisations. The case of the Intergovernmental Panel on Climate Change (IPCC)." *Social Networks* 75:137–147.

Vignola, R., T. L. McDaniels, and R. W. Scholz. 2013. Governance structures for ecosystem-based adaptation: Using policy-network analysis to identify key organizations for bridging information across scales and policy areas. *Environmental Science & Policy* 31:71–84. https://doi.org/10.1016/j.envsci.2013.03.004.

Walker, T. 2023. *Survey: Teachers work more hours per week than other working adults*. National Education Association. https://www.nea.org/nea-today/all-news-articles/survey-teachers-work-more-hours-week-other-working-adults#:~:text=In%20the%20new%20national%20survey,survey%20of%20all%20working%20adults

Wasserman, S., and K, Faust, 1994. *Social network analysis: Methods and applications*. Cambridge, UK: Cambridge University Press.

Weaver, C. P., C. Allen, N. Beller-Simms, T. Fish, A. Grambsch, P. Hipple, M. Kenney, L. Langner, L. Larson, J. S. Lee, L. Marshall, D. McGinnis, S. Mooney, L. Nichols, C. Nierenberg, M. Rockman, A. Rypinski, E. Seyller, and R. Winthrop. 2014. *Social sciences integration to support USGCRP strategic plan implementation: A white paper of the USGCRP Social Sciences Task Force*.

Weaver, C. P., S. Mooney, D. Allen, N. Beller-Simms, T. Fish, A. E. Grambsch, W. Hohenstein, K. Jacobs, M. A. Kenney, M. A. Lane, L. Langner, E. Larson, D. L. McGinnis, R. H. Moss, L. G. Nichols, C. Nierenberg, E. A. Seyller, P. C. Stern, and R. Winthrop. 2014. From global change science to action with social sciences. *Nature Climate Change* 4(8):656–659. https://doi.org/10.1038/nclimate2319.

Weible, C. M., and H. C. Jenkins-Smith. 2016. The advocacy coalition framework: An approach for the comparative analysis of contentious policy issues. In *Contemporary approaches to public policy: Theories, controversies and perspectives*, edited by B. G. Peters, and P. Zittoun. London: Palgrave Macmillan UK. https://doi.org/10.1057/978-1-137-50494-4_2.

Weible, C. M., K. Ingold, D. Nohrstedt, A. D. Henry, and H. C. Jenkins-Smith. 2020. Sharpening advocacy coalitions. *Policy Studies Journal* 48(4):1054–1081. https://doi.org/10.1111/psj.12360.

Weiss, C. H. 1998. Methods for studying programs and policies, 2nd ed. Upper Saddle River, NJ: Prentice-Hall.

West, J. D., and C. T. Bergstrom. 2021. Misinformation in and about science. *Proceedings of the National Academy of Sciences* 118(15):e1912444117. https://doi.org/10.1073/pnas.1912444117.

Whiteman, D. 1985. The fate of policy analysis in congressional decision making: Three types of use in committees. *Western Political Quarterly* 38(2):294–311.

Whyte, K. 2017. Indigenous climate change studies: Indigenizing futures, decolonizing the Anthropocene. *English Language Notes* 55(1):153–162.

Whyte, W. F. 1989. Advancing scientific knowledge through participatory action research. *Sociological Forum* 4(3):367–385. https://doi.org/10.1007/BF01115015.

Wimbush, E., S. Montague, and T. Mulherin. 2012. Applications of contribution analysis to outcome planning and impact evaluation. *Evaluation* 18(3):310–329. https://doi.org/10.1177/1356389012452052.

Worrall, J. L., and E. G. Cohn. 2023. Citation data and analysis: Limitations and shortcomings. *Journal of Contemporary Criminal Justice* 39(3):327–340. https://doi.org/10.1177/10439862231170972.

Yin, R. K. 2009. *Case study research: Design and methods*. Thousand Oaks, CA: Sage.

Zhao, D., and A. Strotmann. 2015. *Analysis and visualization of citation networks*. California: Morgan & Claypool.

Zhao, X., E. Maibach, J. Gandy, J. Witte, H. Cullen, B. A. Klinger, K. E. Rowan, J. Witte, and A. Pyle. 2014. Climate change education through TV weathercasts: Results of a field experiment. *Bulletin of the American Meteorological Society* 95(1):117–130. https://doi.org/10.1175/bams-d-12-00144.1.

Appendix A

Committee Members

Kai N. Lee (*Chair*) is the principal of Owl of Minerva, LLC, a consulting firm working with philanthropic foundations, universities, and nonprofit organizations. Dr. Lee is an adviser on environmental DNA to Oceankind Labs. He is also a fellow of the Center for Ocean Solutions at Stanford University. Dr. Lee is Rosenburg Professor of Environmental Studies, emeritus, at Williams College, and he served as a program officer for science at the David and Lucile Packard Foundation from 2007 to 2015. He is the author of *Compass and Gyroscope* (1993) and co-author, with Howarth and Freudenburg, of *Humans in the Landscape* (2012). At the Packard Foundation, Dr. Lee developed means of targeting funding to achieve near-term practical results in conservation. Earlier in his career, he was a White House Fellow at the U.S. Department of Defense, and a member of the Northwest Power and Conservation Council. Dr. Lee has also served on the Science Advisory Board of the U.S. Environmental Protection Agency. He has served on more than a dozen National Academies of Sciences, Engineering, and Medicine committees and was the founding vice-chair of the National Academies committee to advise the U.S. Global Change Research Program. Dr. Lee served in that role during the committee's review of the third National Climate Assessment, and he was a chapter review editor for the Fourth National Climate Assessment. Dr. Lee holds a degree in experimental physics from Columbia University and a Ph.D. in the same field from Princeton.

Ann M. Gallagher is a science education coordinator for the National Park Service (NPS). She is serving as a governor-appointed commissioner for the Maryland Commission on Parks and Recreation. Other recent experience includes being temporarily embedded in the Climate Change Response Program for the NPS. Gallagher has served for nearly 20 years as the instructor of record teaching undergraduate courses in Earth science and biology. In her federal, nonprofit, and educational research positions, Gallagher navigates between participants, engaged groups, and institutions creating novel science products. For other nonprofit organizations, she has led numerous program and organizational evaluations. Gallagher's areas of expertise include science education, organizational and program management and evaluation, outreach and engagement, watershed and forest ecology, policy drafting and implementation, executive management, and science communication. She holds a graduate certificate in environmental policy from the University of Massachusetts, is an Ecological Society of America senior ecologist, an International Society of Arboriculture certified arborist, and a Project Management Institute program management professional. Gallagher holds an M.S. degree from the University of Maryland and is currently pursuing a Ph.D. in public administration and policy at Old Dominion University's Strome College of Business School of Public Service, uniting climate change science and land management planning.

Matthew Gribble is an environmental epidemiologist and diplomate of the American Board of Toxicology (DABT) currently appointed as the associate chief for research in occupational, environmental & climate medicine at the University of California, San Francisco. He has over a decade of experience in community-engaged and tribal environmental health research on environmental exposures that may contribute to environmental health disparities. Dr. Gribble has provided expert service on topics pertaining to planetary health, climate change and health, and oceans and human health to the World Health Organization, Food and Agriculture Organization of the United Nations, and Intergovernmental Oceanographic Commission. He serves on the National Harmful Algal Blooms Committee and the International Society for Environmental Epidemiology Anti-Racism Task Force. Dr. Gribble is co-editor of *Oceans and Human Health*, a textbook published in 2023. He has previously served as a science advisor to the United States Geological Survey and has received awards for the rigor and creativity of his research from the Universities Council on Water Resources and the Society of Toxicology Metals Specialty Section. In addition to his research program in the United States, Dr. Gribble also has contributed to international climate change adaptation research in Bangladesh. He holds a Ph.D. in epidemiology from Johns Hopkins University.

Scott Kalafatis is the deputy university director of the Northwest Climate Adaptation Science Center affiliated with the Climate Impacts Group at the University of Washington. Previously, Scott was an assistant professor of sustainability at Chatham University, a visiting assistant professor of environmental studies at Dickinson College, a National Science Foundation ethical STEM postdoctoral researcher with the College of Menominee Nation, and a researcher with Great Lakes Integrated Sciences and Assessments. Dr. Kalafatis is a social scientist with a research background focused on understanding collaborations between scientists and decision-makers, including how these collaborations contribute to decisions informed by climate science, what the ethical implications of these collaborations are, and how to train and support scientists pursuing them. He has leveraged training methods and research design strategies from across a variety of social science and applied fields to evaluate the use of climate science in decisions made by local governments, tribal governments, and resource managers throughout the United States. Dr. Kalafatis holds M.A.s in environmental policy and urban planning, and a Ph.D. in resource policy and behavior from the University of Michigan.

Jessica Kronstadt is the program director at the Planetary Health Alliance, a global consortium of organizations committed to understanding and addressing the impacts of environmental change on human health and well-being. Previously, she was the vice president for research and product development at the Public Health Accreditation Board (PHAB) where she led PHAB's research and evaluation portfolio as well as the development of the most recent version of the national, consensus-based accreditation standards for state, tribal, local, and territorial health departments. Throughout her career, Kronstadt has led or advised numerous evaluations of public health and other health initiatives. For more than a decade, she led evaluation activities to determine the impact of health department accreditation. Her other evaluations of complex programs include the U.S. Center for Disease Control and Prevention's Public Health Infrastructure Grant and the Health Information Technology and Economic and Clinical Health (HITECH) Act workforce development program. Kronstadt has served as chair of both the Montgomery County, Maryland, Commission on Health and the Academy Health Public Health Systems Research Interest Group. She received her M.P.P. from Georgetown University and her B.A. from Amherst College.

Glynis Lough is an affiliate at the Aspen Global Change Institute (AGCI), where she also directs two collaboratives: the Practitioner Exchange for Effective Response to Sea Level Rise and the Science for Climate Action Network. Previous positions have included vice president, grants for the National Geographic Society; deputy director and research director, Food & Environment Program of the Union of Concerned Scientists; chief of staff for the Third National Climate Assessment with the U.S. Global Change Research Program; and American Association for the Advancement of Science and Technology Policy Fellow in the U.S. Environmental Protection Agency Office of Air, Office of Policy Analysis and Review. Dr. Lough works with AGCI to advance collaborative learning processes that accelerate climate solutions, supporting effective co-development of practical, vetted, real-world knowledge. Throughout her work, she seeks to solve complicated issues by centering humans' lived experiences and co-developing knowledge to achieve meaningful outcomes. With a background in engineering, Dr. Lough brings a wealth of experience in air quality, climate change, food systems, social justice, public health, and conservation. She holds a Ph.D. in environmental chemistry and technology from the University of Wisconsin–Madison and a B.S. in chemical engineering from Case Western Reserve University.

Michelle Miro is a senior information scientist at the RAND Corporation and professor at the Pardee RAND Graduate School. She has a broad portfolio of work across climate resilience and adaptation and disaster recovery for infrastructure, with a focus on the water sector. Dr. Miro's research supports international, federal, and local emergency, infrastructure, and resource management agencies with climate adaptation, disaster resilience and recovery, and water resources planning. Her recent projects have developed national and regional climate risk assessments of U.S. critical infrastructure systems, as well as climate services that support local to regional entities interpreting and acting on climate information. Dr. Miro holds an M.S. in civil engineering from the University of Illinois Urbana-Champaign and a Ph.D. in civil engineering from the University of California, Los Angeles.

Ariane Pinson is the research social scientist for the U.S. Army Corps of Engineers (USACE), Albuquerque District. She conducts climate resilience planning for both civil and military projects, advises on climate policy and guidance for USACE and the Department of the Army, and develops data and decision support tools to support climate resilience planning across the Department of Defense. Dr. Pinson has a sustained research focus on climate change impacts to landscapes, ecosystems, and human adaptation and serves as the federal coordinating lead for the water chapter for the Fifth National Climate Assessment. She holds a Ph.D. in anthropology from the University of New Mexico and a B.A. in geography.

Urooj Raja is an assistant professor at the School of Communication at Loyola University Chicago in the Department of Advocacy and Social Change where her focus is on the human dimensions of environmental problems. Dr. Raja is an interdisciplinary social scientist whose work focuses on how the general public engages with and responds to climate change. Her recent work has been published in *Scientific Reports* and *Environmental Communication*. Dr. Raja is currently a review editor for the journal *Frontiers: Science and Environmental Communication* and sits on the Executive Section of the American Geophysical Union's Science and Society Section and the Environmental Communication section of the National Communication Association. She has previous work experience in the field of environmental philanthropy (the Solutions Project), polling behavior (Pew Research Center), and local and national decision-making bodies (New York State Assembly and the U.S. House of Representatives). Dr. Raja holds a Ph.D. in environmental studies and an M.A. in environmental sociology from the University of Colorado Boulder and a B.A. in history from Princeton University.

Carlos Rodriguez-Franco is a senior forester with the Research and Development arm of the U.S. Forest Service and served as the deputy chief from 2016 to 2018. Dr. Rodriguez-Franco worked for Mexico's National Institute of Forestry, Agriculture and Animal Husbandry Research for 25 years and served as the general director from 1996 to 2000. His work focuses on forest inventories, silviculture, forest management, carbon sequestration, biochar, plant production techniques, forest plantations, and agroforestry systems. Dr. Rodriguez-Franco has taken leadership training at the John F. Kennedy School of Government and with Dialogos. He was awarded the category of national researcher by the National System of Researchers in Mexico from 1987 to 2000 for conducting outstanding research, the results of which have benefited Mexican technology and have contributed to increased productivity and improved basic knowledge. Dr. Rodriguez-Franco holds a Ph.D. in forest sciences from Yale University.

Kathleen Segerson is a Board of Trustees Distinguished Professor of Economics at the University of Connecticut. She is an environmental economist, with a strong interest in collaborative interdisciplinary work. Dr. Segerson's research within economics is primarily applied theory focused on the incentive effects of alternative environmental and conservation policy instruments, with applications to groundwater contamination, hazardous waste management, land use regulation, climate change, nonpoint pollution from agriculture, and protection of marine species. She is a member of the National Academies of Sciences, Engineering, and Medicine and a Fellow of the Association of Environmental and Resource Economists, the Agricultural and Applied Economics Association, and a fellow at the Beijer Institute of Ecological Economics in Stockholm. Dr. Segerson has held numerous editorial positions and has served on a number of advisory boards, including the U.S. Environmental Protection Agency's Science Advisory Board and the Committee on Valuing the Protection of Ecological Systems and Services. She has served as a member of the National Academy of Sciences Advisory Committee for the U.S. Global Change Research Program and on the review panel for the Third National Climate Assessment. Dr. Segerson holds a Ph.D. from Cornell University and a B.A. from Dartmouth College.

Kristin Timm is a research assistant professor at the International Arctic Research Center at the University of Alaska Fairbanks (UAF). Dr. Timm's expertise is in science and climate change communication and the people and processes at the interface of science and society. She is a member of the Alaska Climate Adaptation Science Center leadership team, directing research and activities related to communication and actionable science. Prior to her time at UAF, Dr. Timm worked and studied with the Center for Climate Change Communication at George Mason University and spent over a decade working as a science education project manager and professional science communicator in Alaska. Her dissertation investigated news coverage of the Fourth National Climate Assessment and the factors that influenced it. Dr. Timm has received several awards for her work, including the U.S. Geological Survey Eugene M. Shoemaker Communication Award for effectiveness communicating complex scientific concepts. She previously served as a member of the National Academies of Sciences, Engineering, and Medicine committee to review the Fifth National Climate Assessment. Dr. Timm has a Ph.D. in communication from George Mason University and a M.Sc. in interdisciplinary studies and a B.A. in rural development: land, resources, and environmental management, both from the University of Alaska Fairbanks.

Appendix B

NCANet

NCANet Participants as of 2021[1]

- Acclimatise
- Adaptation International
- Adaptation Now
- Agrivarsity Scouts Group
- Alaska Ocean Observing System
- Alliance for Climate Education
- Alpha Epsilon Lambda
- American Association for the Advancement of Science Center for Public Engagement
- American College and University Presidents' Climate Commitment
- American Geophysical Union
- American Geosciences Institute
- American Institute of Physics
- American Lung Association
- American Meteorological Society Policy Program
- American Planning Association
- American Public Health Association
- American Public Health Association Center for Climate, Health, and Equity
- American Society of Adaptation Professionals
- American Society of Civil Engineers
- American Statistical Association
- American Water Works Association
- Animalia Project
- Association for the Tree of Life
- Association of Climate Change Officers
- Association of Fish & Wildlife Agencies

---

[1] Archive of http://ncanet.usgcrp.gov/partners via the Internet Archive Wayback Machine https://web.archive.org/web/20210913184951/http://ncanet.usgcrp.gov/partners (accessed August 9, 2024).

- Association of Metropolitan Water Agencies
- Association of Natural Resource Extension Professionals
- Association of Public Health Laboratories
- Association of State and Territorial Health Officials
- Association of Zoos and Aquariums
- BCC Planning
- Bear Path Acres Animal Education Center
- better, Inc.
- Biositu, LLC
- Blue Frontier Campaign
- BlueGreen Alliance
- Broward County (Florida)
- The Cadmus Group
- California Ocean Science Trust
- California State University East Bay
- Campus Climate Corps
- Carnegie Mellon University Engineering and Public Policy Program
- Carolinas Integrated Sciences and Assessments (CISA)
- Carpe Diem West
- CASE Consultants International
- Center for Clean Air Policy (CCAP)
- Center for Climate and Energy Solutions
- Center for Climate and Security
- Center for Integrated Solutions to Climate Challenges, Arizona State University
- Center for Research on Environmental Decisions, Columbia University
- Center for Sustainable Cities, University of Southern California
- Center for Urban Real Estate, Columbia University
- CEPT University
- Chamber of Eco Commerce
- Chicago Metropolitan Agency for Planning
- Chicago Zoological Society
- Citizens Climate Lobby
- City of Cambridge Community Development Department
- The CLEO Institute
- Climate Access
- Climate Assessment for the Southwest (CLIMAS)
- Climate Central
- Climate Change & Society Master's Program, North Carolina State University
- Climate Change Institute, University of Maine
- Climate Communication
- Climate Data Solutions, LLC
- Climate Generation: A Will Steger Legacy
- Climate Impacts Group, University of Washington
- Climate Literacy and Energy Awareness Network (CLEAN)
- Climate Literacy Partnership in the Southeast (CLiPSE)
- Climate Nexus
- Climate Realty, LLC
- Climate Resolve
- Climate Science and Policy Watch
- Climate XChange

APPENDIX B

- CMC Corporate Solutions
- Coastal Protection and Restoration Authority—Louisiana
- CoClimate
- Collaboratory for Adaptation to Climate Change
- Conservation Biology Institute
- Consortium for Ocean Leadership
- Consortium for Science, Policy, and Outcomes—Arizona State University
- Consortium of Universities for the Advancement of Hydrologic Science, Inc.
- Cooperative Institute for Climate and Satellites—North Carolina (CICS-NC)
- Council of State and Territorial Epidemiologists Climate Change Workgroup
- Crop Science Society of America
- CU (University of Colorado) Sea Level Research Group
- Dahl Scientific
- Dan River Basin Association
- Defenders of Wildlife
- Delta Land & Community
- Drawing Conclusions LLC
- Ducks Unlimited, Inc.
- Earth to Sky Interagency Partnership
- East-West Center
- EcoAdapt
- ecoAmerica
- Ecological Society of America
- Ecologic Institute
- Electric Power Research Institute
- Energy/Climate AAAS Fellows and Alumni
- Environmental and Energy Study Institute
- ESIP Federation
- Florida Climate Institute
- Four Twenty-Seven, Inc.
- Friends of Blackwater
- Geological Society of America
- Geos Institute
- Government Accountability Project
- Great Lakes Indian Fish and Wildlife Commission
- Great Lakes Integrated Sciences and Assessments (GLISA)
- Great Plains Institute
- Greater Washington Interfaith Power and Light
- Green City Environment Group
- GUND Institute for Ecological Economics
- Habitat Seven
- Harte Research Institute for Gulf of Mexico Studies
- IAPMO
- Icahn School of Medicine at Mount Sinai
- ICF International
- ICLEI—Local Governments for Sustainability USA
- Institute at Brown for Environment & Society (IBES)
- Institute of Caribbean Studies
- Institute for Coastal Science and Policy, East Carolina University
- Intelligentsia International

- Inter-American Development Bank
- Internews Earth Journalism Network
- iOffice
- IOOS Association
- Island Press
- Kansas State University
- The Keystone Center
- Kleinschmidt Associates
- L & R Resources, LLC
- Labor Network for Sustainability
- Legacy Pathways LLC
- Livelihoods Knowledge Exchange Network
- LMI
- Local Government Commission
- Macconit
- Madison River Group
- Manzanita Tribe
- Maryland and Delaware Climate Change Education Assessment and Research (MADE CLEAR)
- Massachusetts Climate Action Network
- Metcalf Institute for Marine & Environmental Reporting
- Metropolitan Council
- Milken Institute School of Public Health
- Mississippi-Alabama Sea Grant Consortium
- Musial Associates, LLC
- National Academy of Engineering Center for Engineering, Ethics, and Society
- National Association of County and City Health Officials
- National Center for Science Education
- National Council for Science and the Environment
- National Ecological Observatory Network, Inc.
- National Environmental Health Association
- National Wildlife Federation
- Natural Resources Defense Council
- NC Interfaith Power & Light
- Nebraska Climate Change Assessment
- New England Aquarium
- New York City Department of Environmental Protection
- New York Presidents for Climate Action (NYPCA)
- Next Generation
- North Carolina Department of Health and Human Services
- North Central Climate Science Center
- Northeast Climate Science Center
- Oregon Department of Agriculture
- Oregon Health Authority WIC
- Pacific Islands Regional Climate Assessment (PIRCA)
- Pacific Northwest Tribal Climate Change Network
- Pacific Regional Integrated Sciences and Assessments (Pacific RISA)
- Panattoni Development Company, Inc.
- Philadelphia Water Department
- Playa Lakes Joint Venture
- Portland Water Bureau

APPENDIX B

- Program on Climate and Health, Center for Climate Change Communication
- Public Health Institute—US Climate and Health Alliance
- Puget Sound Clean Air Agency
- Purdue Climate Change Research Center
- The Rainmakers—TBL
- Resource Media
- Rising Voices
- Running Start Institute
- Rutgers Climate Institute
- San Diego County Regional Airport Authority
- Saving the Planet—An Ecology Game
- The Science and Resilience Institute at Jamaica Bay
- Second Nature
- Security and Sustainability Forum
- Sierra Club
- Sierra Club—Michigan Chapter
- Society of American Indian Government Employees (SAIGE)—USDA Chapter
- Soil Science Society of America
- South Central Climate Science Center
- Southeast Climate Consortium
- Southern Climate Impacts Planning Program
- Stanford Woods Institute for the Environment
- State Climate Office of North Carolina
- Sustainable Northwest
- Sustainable Rangelands Roundtable
- Sustain NC
- TERC
- Texas Sea Grant
- Union of Concerned Scientists
- United South and Eastern Tribes, Inc.
- University Corporation for Atmospheric Research
- University of Alabama in Huntsville, College of Nursing
- University of Arizona Center for Climate Adaptation Science and Solutions
- University of California, Berkeley Department of Environmental Science, Policy, and Management
- University of South Carolina Environment & Sustainability Program
- US CLIVAR
- USA National Phenology Network
- Ute Mountain Ute Tribe
- Value Sustainability
- Ward Circle Strategies
- Water Utility Climate Alliance
- Weather Blogger
- WeSpire
- Western Water Assessment
- Willis Towers Watson
- World Institute on Disability
- World Resources Institute
- World Wildlife Fund—US
- Yale Climate and Energy Institute
- Young Voices for the Planet

U.S. Government Participants

- Florida Sea Grant
- NASA Goddard Space Flight Center
- National Park Service
- Southwestern Regional Climate Hub
- U.S. Bureau of Reclamation
- U.S. Environmental Protection Agency Office of Water
- U.S. General Services Administration
- USDA APHIS
- USDA Climate Hubs
- USGCRP Adaptation Science Working Group
- USGCRP Climate Change and Human Health Working Group

Appendix C

Crosswalk Between Statement of Task and Overarching Evaluation Questions

Table C-1 below provides a map connecting the themes and questions provided in the Statement of Task to the preliminary overarching evaluation questions presented in Chapter 3. As can be seen in the table, all of the questions that appear in the Statement of Task have one or more corresponding overarching evaluation questions or subquestions. It is also important to note that while there is very close alignment between the Statement of Task questions and the preliminary overarching evaluation questions, there are some differences in framing. This reflects the committee's decision to align the evaluation questions with the preliminary logic model. Throughout the evaluation design process carried out by USGCRP and their consultant, the committee anticipates that these questions will continue to evolve as a more refined logic model is developed that reflects an updated understanding of the program and the priorities of the key evaluation users. There are several overarching evaluation questions or subquestions presented in this report that do not correspond to the questions in the Statement of Task. Specifically, the committee added question 5 ("How do the contextual factors described in the logic model influence how audiences feel, what they gain (e.g., cognition), what they do (e.g., behaviors), and how do they mediate the use of the NCA?") and subquestion 5a ("How did the characteristics of the audience—including those who have been historically disenfranchised—affect the extent to which their informational and decision-making needs were met?"). This decision reflects the emphasis in the Statement of Task that the committee consider diversity, equity, inclusion, and justice (DEIJ) principles.

In addition, as described in Chapter 4, the committee finds that exploring the network of networks approach is critical to understanding the mechanisms through which the NCA and other USGCRP products are used. As such, the committee introduced evaluation question 6 ("How does the network of networks factor into use?") This will illuminate the Statement of Task themes with regards to usefulness, decision making, and awareness/engagement. However, because the committee's framing is specific to the network of networks concept, those questions are not presented in the table.

TABLE C-1 Crosswalk Between Themes and Questions from the Committee's Statement of Task and Preliminary Overarching Evaluation Questions Presented in Chapter 3

Statement of Task Themes	Statement of Task Questions	Preliminary Overarching Evaluation Questions or Subquestions
Regarding usefulness	How and to what extent have stakeholders found NCA materials to be useful? What specifically has been useful?	• How, and to what extent, did NCA products address information needs among priority audiences (i.e., what did they gain cognitively in terms of knowledge, skills, attitudes, capacities, etc.)? (Q2) • Which products or parts of products contributed to meeting information needs? (Q2b) • How, and to what extent, did NCA products address decision needs among priority audiences (i.e., what did they do as a result of using the products)? (Q3) • Which products or parts of products contributed to meeting decision needs? (Q3b) • How did the characteristics of the audience—including those who have been historically disenfranchised—affect the extent to which their informational and decision-making needs were met? (Q5a)
	Does the selection of topics, regions, sectors, and level of detail (e.g., time frame, spatial resolution, subsector concerns) in the NCA (and USGCRP products it references as well as related USGCRP products) adequately address the stakeholders' needs?	• How does the selection of topics, regions, sectors, level of detail, ways of expressing uncertainties, and other aspects of what information is included and the way it is organized and presented contribute to audiences' perceptions of how the products meet their needs? (Q4b)
	What decision or informational needs were well-addressed by the NCA? What decisions can stakeholders make given the level of information provided?	• What informational and other cognitive needs were well addressed by the NCA? (Q2a) • What decision needs were well addressed by the NCA? What decisions can audiences make given the level of information provided? (Q3a)
Regarding decision-making, future needs, and missing information, details, and/or tools	What future needs are anticipated? What additional types of decisions (if any) do stakeholders anticipate they would revisit given different topics and/or levels of detail? What information would be required to meet needs that USGCRP is not meeting already?	• Given the evolving nature of the field, what are future needs and how could USGCRP help meet them? (Q5c)
	What decision or information needs did the stakeholders expect would be met by the NCA but were not?	• What wasn't gained that could or should have been? What are the gaps in what the NCA provided and what else did potential audiences use to cover them (including sources based on USGCRP products)? Are these gaps that audiences expect the NCA to meet? (Q2c) • What actions weren't taken because audiences did not have their decision needs adequately met? What are the gaps in what the NCA provided and what else did potential audiences use to cover the gaps or for other purposes (including sources based on USGCRP products)? Are these gaps that audiences expect the NCA to meet? (Q3c) • Given the abundance of other climate change resources, what is the unique value of the NCA and related products? (Q4c) • How did the context of multiple information sources affect the impact of NCA? Did they expand its impact? Did they replace it? (Q5b)

APPENDIX C

TABLE C-1 Continued

Statement of Task Themes	Statement of Task Questions	Preliminary Overarching Evaluation Questions or Subquestions
	For stakeholders whose decision or informational needs were not met by the NCA and selected USGCRP products, what is the reason? What other products/materials, including other USGCRP and non-USGCRP products, did they use, if any?	• What wasn't gained that could or should have been? What are the gaps in what the NCA provided and what else did potential audiences use to cover the gaps or for other purposes (including sources based on USGCRP products)? Are these gaps that audiences expect the NCA to fill? (Q2c) • What actions weren't taken because audiences didn't have the info that was needed? What are the gaps in what the NCA provided and what else did potential audiences use to cover the gaps or for other purposes (including sources based on USGCRP products)? Are these gaps that audiences expect the NCA to meet? (Q3c)
Stakeholder awareness and engagement	How aware or involved were different stakeholder groups in the NCA development process and how did this influence their use of the report? For stakeholders who were not previously aware of the NCA development process, how did they become aware of the NCA?	• To what extent are priority audiences aware of NCA products and what are the most effective ways to increase awareness? Did involvement in the development process contribute to general awareness and use of the report? (Q1) • For audiences who were not previously aware of the NCA, how did they become aware of the NCA? What factors, including the NCA development process, are related to awareness of the NCA? (Q1a) • How, and to what extent, did users share information from NCA products with others? What could be done to make it easier for users to push information out to others? (Q1b) • To what extent are audiences using other resources that are based on USGCRP products? (Q1c) • What groups have been involved in the NCA development process and to what extent were they involved? How effectively were historically marginalized and underrepresented stakeholders included? (Q1d) • How did engagement in the process affect use of the NCA? How did the engagement of historically marginalized communities and underrepresented audiences affect use of the NCA? (Q4a)
	How effectively does the NCA development process engage historically marginalized communities and underrepresented stakeholders?	• How did engagement in the process affect use of the NCA? How did the engagement of historically marginalized communities and underrepresented audiences affect use of the NCA? (Q4a) • How did the characteristics of the audience—including those who have been historically disenfranchised—affect the extent to which their informational and decision-making needs were met? (Q5a)
	How understandable and navigable is the NCA, including the report documents and findings, and underlying supporting data? Is the NCA information presented in a format that informs decision-making?	• How did the attributes of the products and process contribute to how users feel, what they gain (e.g., cognition), and what they do (e.g., behaviors)? What about the products and process could be changed to make them more effective? (Q4) • How does the selection of topics, regions, sectors, level of detail, ways of expressing uncertainties, and other aspects of what information is included and the way it is organized and presented contribute to audiences' perceptions of how the products meet their needs? (Q4b)

continued

TABLE C-1 Continued

Statement of Task Themes	Statement of Task Questions	Preliminary Overarching Evaluation Questions or Subquestions
Items added based on the logic model		• What organizations or types of organizations are the most powerful in spreading USGCRP products to other audiences (i.e., most influential nodes in the network)? What are the approaches that best support the sharing of information through the network of networks? (Q6a) • How does USGCRP instruct or amplify the work that is being done by agencies and other partner organizations? How does the information generated by the NCA help facilitate their work? (Q6b) • Which part of this effort directly involves USGCRP vs what is appropriate for the extended network? (Q6c) • How did NCA affect other information sources regarding climate change? In other words, do those information sources make use of it? (Q6d) • Has NCA affected the degree to which researchers and decision-makers collaborate? How does NCA affect the identification of research needs or development of additional research? (Q6e) • Who is left out in the network of networks approach? (Q6f) • How do frontline federal workers in these agencies feel, what do they gain, what do they do as a result of the NCA? (Q6g)

Appendix D

Network Analysis: Additional Measures and Strategies

Network analysis features concepts and techniques that could potentially be of value to those involved with the evaluation of the use of NCA products. In the process of developing the approaches used across many different contexts and applications, however, network analysts have created a great deal of technical terminology and methodology specific to their perspectives on networks. These terms and methods are not widely familiar, and that can limit recognition of opportunities to use network analysis within an evaluation, complicating communication with network analysis experts. Chapter 4 provides an overview of how network analysis could support exploration of a network of networks associated with the use of NCA products. This appendix provides additional context about network analysis measures and strategies that are anticipated to be relevant for addressing the type of evaluation discussed in this report. It is meant to aid those involved in the evaluation who do not have backgrounds in network analysis with making sense of how network analysis might support their efforts and how to communicate with network analysts.

A basic distinction that network analysts commonly make is between different types of networks. The most common types are egocentric, monopartite, and bipartite networks. An egocentric network focuses on a particular node at the center with connections extending from it, such as asking an individual who they collaborate with the most and then mapping the network based on those individuals and their links to other nodes branching off from them. Such an approach might be used with USGCRP at the network's center to develop a map of the network of networks as perceived by the program.

Most network maps do not start with a central node at the outset and instead focus on the larger system. Monopartite networks have only one kind of node, such as a network where peer-reviewed publications are the nodes and links are made between them based on which papers cite one another. It is also common to use bipartite networks, which have two kinds of nodes. If the citation network was approached as a bipartite network, it could include both the peer-reviewed publications and the people who authored them. In that network map, authors would be linked with particular papers and then the papers would be linked with one another. Event data are also commonly mapped as a bipartite network where participants are linked with particular events. Monopartite and bipartite networks are also commonly referred to as one-mode and two-mode networks, respectively.

As mentioned in Chapter 4, there are different ways network analysts assess the centrality of nodes based on how they want to understand how the location of nodes might affect outcomes of interest. Understanding which node has the most total connections can be an indicator of importance, and *degree* describes the number of links a particular node has with the other nodes in a network. However, a node might have many direct ties with a particular set of links in a network, but then be relatively isolated from many other parts of the network. *Closeness* addresses this potential isolation relative to all the other nodes in the network by calculating the average number of links that need to be traversed to get from any particular node to the other nodes in the network. Another consideration is the influence a node ultimately has on the network, which might be determined by examining the influence of the other directly linked nodes. Eigenvector centrality ranks nodes based on their connection with highly ranked nodes. Google's PageRank algorithm, which allowed its original search engine to innovatively identify and promote higher-quality webpage results, is an adaptation of eigenvector centrality (Brin and Page, 1998). Sometimes the importance of nodes in a network is not about the number of connections, but about how critical the node's position is for connecting parts of the network that are isolated from one another. *Betweenness* describes the number of times a particular node is a part of the shortest route between two other nodes in the network. Identifying such key connections between parts of a network that are otherwise disconnected can be a powerful application of network analysis. The importance of those who bridge between structural holes in networks, described in Chapter 4, is associated with the recognition of the importance of nodes with high betweenness.

Chapter 4 highlights citation analysis as an established application of network analysis that can support an evaluation, but there are different ways that it can be approached depending on what is trying to be understood. Different types of relationships within these citation networks may be of interest, with common distinctions made between patterns of direct citation (one thing cites another), co-citation (what things are cited at the same time), and bibliographic coupling (multiple entities citing the same thing; Boyack and Klavans, 2010). These three techniques provide options for focusing on different aspects of these networks: direct citation highlights immediate connections between information sources and use; co-citation highlights associations between similar information sources; and bibliographic coupling highlights associations among similar information consumers. Researchers have also developed a variety of techniques to enhance the nuance and insight that can be gained from citation analysis by accounting for the context surrounding citations themselves. Content-based citation analyses often include automated techniques that address and account for things such as motivations underlying citation and function a citation serves in a document (for an introduction, see Ding et al., 2014). Such techniques are valuable because citations serve at least five different types of functions for authors, and there is still debate in the field even about the extent to which citations themselves should be treated as indications of intellectual connections or social connections (Worrall and Cohn, 2023).

When used effectively, these methods can help address common limitations of citation analysis associated with a lack of nuance about why the citations appear. However, there are still several other common limitations that those involved in evaluation should consider (Worrall and Cohn, 2023). Focusing on citations narrows the scope of analysis to specific kinds of uses of information that can be captured consistently in citations and can limit the scope of assessment to only certain kinds of documents, namely those freely available online in the languages included in the analysis and indexed in a way that makes them identifiable. Inclusion of non-academic publications in citation analysis can help provide insight into non-academic impacts of information; however, including them substantially increases the time and labor necessary to perform analysis, as identification of these documents cannot currently be automated and their quality and content are less consistent than academic publications (Sibbald et al., 2015). Including these types of documents also does not fully address equity concerns about inclusion in evaluation because it leaves out audiences who do not produce documents like these or do not post them online. The study of weblinks is based on citation analysis (Thelwall, 2012), and links can be viewed as web citations akin to citations between documents (Dudek et al., 2021). However, researchers have to account for the motivations surrounding website links, as they differ from the motivations behind academic citations (Björneborn and Ingwersen, 2004).

Chapter 4 also highlights the study of social media networks, but those involved with evaluation might also benefit from additional perspectives about how this type of data can be used. Two common types of applications of social network analysis and social media are *embedding learning* and *community learning* (for a review, see Bazzaz Abkenar et al., 2021). Embedding learning models aims to determine how different nodes within social networks influence information diffusion based on a user's characteristics and how they disseminate information. Community learning models are focused on identifying clusters of users who represent people with similar characteristics, interests, or attitudes.

Network data about event participation is also considered to be relatively easy to obtain through taking attendance or drawing it from conference call agendas or participant lists. These data can be connected with other data sources such as surveys about contacts' close colleagues. It is still common to collect network data through questions in surveys or interviews. If the participants in the network are known in advance, it is common to develop the network map based on asking them about their relationships with one another and outcomes of interest. If the extent of the network is not known in advance, snowball sampling techniques are often used where respondents are asked to suggest subsequent respondents. For example, Cunningham et al. (2015) were able to eventually reach residents associated with the spread of climate adaptation information through a snowball sampling method that included government officials and community-based organizations. Cvitanovic et al. (2017) assessed the expanding impact of a particular knowledge broker by having them report their egocentric network every 3 months throughout the year. The broker's perception of the strength and effectiveness of their connections was compared with that of surveyed participants in the network to assess changes in the strength of relationships over the year as well. Masuda et al. (2018) also created a research design based on known and unknown participants to assess diffusion. Working with The Nature Conservancy, they tested how sharing organizational learning workshop links with different types of people affected participation in the workshops and demonstrated that using informal contacts could enhance diffusion and changes in attitude.

Appendix E

Derivative Products

It is important to bear in mind that climate information traceable to the NCA and its associated products sometimes appears in derivative products. Most of the entities and populations likely to use the NCA are reached only indirectly, many getting information via products and services prepared outside the NCA process but making use of its information.

Subnational governments, including state, local, tribal, and territorial governments, use the information gathered in the NCA to make planning decisions. Numerous federal programs, grants, service programs, and partnership programs support these entities by providing observations of weather and climate, as well as downscaled extreme weather and climate model data, much of which draws directly upon the data used to support the NCA. An example of a planning decision is the effort of the City of Miami Beach, Florida, to increase resilience to rising sea levels through road elevation and other actions.[1]

Nongovernmental organizations and businesses use the NCA and underlying data to provide information and make decisions. Most of these audiences will combine climate data with other information in the decision-making process. For example, the Water Utility Climate Alliance serves as a clearinghouse for information that can assist water utilities to plan for climate change. They provide guidance, educational programs, and advocacy that draws on information from the NCA and its underlying data.

Educators rely primarily on derivative products—textbooks, podcasts, and literary works—to bring knowledge of climate change impacts, adaptation, and mitigation to the classroom. As part of their educational mission, some federal agencies provide materials that closely align with existing state or other standards to educators. For example, the NASA Jet Propulsion Lab has created K–12 climate change lessons and activities aligned to Next Generation Science and Common Core Math Standards.[2] Similarly, the National Park Service is a source of informal science and climate education to over 325 million visitors each year (Campbell et al., 2020).

---

[1] See https://www.miamidade.gov/global/economy/resilience/sea-level-rise-flooding.page.
[2] See https://www.jpl.nasa.gov/edu/teach/tag/search/Climate+Change.

The news media, including weather reports for television news, draw upon such organizations as Covering Climate Now,[3] World Weather Attribution,[4] and Climate Central.[5] These communications entities synthesize climate change information for a nontechnical audience and draw upon the NCA and associated products.

In these and other cases, information from the NCA can be used to create derivative products and services. As the evaluation explores pathways highlighted in the logic model, the evaluation design should take into account the way that derivative products provide decision support by distributing selected information from the NCA.

---

[3] See https://coveringclimatenow.org/.
[4] See https://www.worldweatherattribution.org/.
[5] See https://www.climatecentral.org/.

Appendix F

Further Information About Federal Programs

As noted in the text above, there are several federal programs that are potential users of the NCA and its products. These include:

- The National Oceanic and Atmospheric Administration (NOAA) Climate Adaptation Partnerships (CAP)/Regional Integrated Sciences and Assessments program consists of 14 centers across the United States and territories focusing on applied climate change research with a regional focus and emphasizing community collaboration and partnerships. The focus is on regional capacity to adapt to climate impacts.[1] These are NOAA-led.
- The NOAA National Centers for Environmental Information Regional Climate Center Program[2] provides existing weather extremes and climate data, as well as climate change-related research and services. These are led by the lead academic partner.
- The U.S. Geological Survey (USGS) has 10 (9 regional and 1 national) Climate Adaptation Science Centers for the purpose of developing science, data, and tools to help natural and cultural resource managers address the impacts of climate change on fish, wildlife, ecosystems, and the communities they support. A key part of the effort is the creation of datasets, web applications, assessments, surveys, and other tools that are then made publicly available for future management or research projects. These are hosted at academic institutions and co-led by the USGS and host institution, and they receive guidance from regional participants and audiences, including tribes.[3]
- The U.S. Fish and Wildlife Service is the lead agency for the 22 National Landscape Conservation Cooperatives (LCCs), an organization that includes federal, state, local, and NGO participants. These are meant to function as applied science and ecosystem management collaborations. The LCCs focus on habitat resilience to large landscape-level stressors and those stressors that may be magnified by the changing climate.[4]

---

[1] See https://cpo.noaa.gov/divisions-programs/climate-and-societal-interactions/cap-risa/.
[2] See https://www.ncei.noaa.gov/regional/regional-climate-centers.
[3] See https://www.usgs.gov/programs/climate-adaptation-science-centers/science/science-tools-managers.
[4] See https://www.landscapepartnership.org/cooperative.

- The U.S. Department of Energy leads the Grid Resilience Technical Assistance Consortium, which is a national-scale center to support state energy officials, public service commissions, and utilities through the analysis of climate change threats, expected impacts, best practices for adaptation and investment, and resilience planning.[5]
- The U.S. Department of Health and Human Services has organized an Office of Climate Change and Health Equity,[6] which relies on USGCRP products such as "The Impacts of Climate Change on Human Health in the United States: A Scientific Assessment."[7]

All of these organizations interact with and share data with each other and with other federal agencies, forming a large network of networks. The following provides additional descriptive information about the U.S. Department of Agriculture (USDA) climate hubs, which are used in the text as an example of a federal program that might be included in a USGCRP evaluation of the use of the NCA and its products.

USDA CLIMATE HUBS

The USDA climate hubs are a collaboration across the department's agencies that are led and hosted by the Agricultural Research Service, Forest Service, and Natural Resources Conservation Service (USDA, 2023b). These 11 hubs develop and deliver science-based, region-specific information and technologies to enable climate-informed decision-making. Each Regional Climate Hub is based out of a Forest Service research station or Agricultural Research Service lab. The International Climate Hub is based out of the Foreign Agricultural Service. An executive committee of the climate hubs includes the above agencies as well as the Economic Research Service, Farm Service Agency, National Institute of Food and Agriculture, Office of Energy and Environmental Policy, Risk Management Agency, and Rural Development. Each hub is tasked with addressing the unique climate challenges and opportunities of its region, while also working with other hubs to develop and share information nationwide (USDA, 2021).

The climate hubs provide periodic regional assessments of risk and vulnerability to production sectors and rural economies, building on material provided through the NCA. The hubs use existing climate change information and assessments to identify primary risks to producers from increasing weather variability and a changing climate. These assessments/syntheses consist of information on vulnerabilities of primary regional agricultural, ranching, and forestry commodities to climate change effects and identify mitigation and adaptation strategies (USDA, 2021). The climate hubs draw upon the NCA's regional chapters when addressing the hubs' needs for more detailed information.

The climate hubs create a dialogue between the services the agency is authorized to provide and the needs expressed by individuals and the communities they serve. The inward dialog allows the climate adaptation needs of the agriculture and ranching communities to inform the direction of USDA support. Employees at the hubs who cover the needs of USDA customers and the public in general focus their work using NCA information to cover eight priority areas: (1) adaptation/resilience, (2) mitigation, (3) adaptation plan implementation, (4) wildland fire management/restoration, (5) climate literacy, (6) climate-smart agriculture and forestry practices, (7) environmental justice, and (8) international work. Hubs further operate through three main relevant work areas which often draw on NCA scientific information (NOAA, 2024c; USDA, 2023b). These three areas are discussed below.

1. **Science and Data Synthesis**: The hubs translate and deliver the latest NCA climate science information relevant to users. This includes regional risk and vulnerability assessments[8] as well as contributions to the National Climate Assessment.
2. **Tool Development and Support**: The hubs help develop and support tools that enable climate-informed planning and decision-making. So far, the hubs have collaborated on the development of over 25 climate-based tools.[9]

---

[5] See https://www.energy.gov/gdo/grid-resilience-technical-assistance-consortium.
[6] See https://www.hhs.gov/climate-change-health-equity-environmental-justice/climate-change-health-equity/index.html.
[7] See https://health2016.globalchange.gov/.
[8] See https://www.climatehubs.usda.gov/hubs/northwest/topic/northwest-vulnerability-assessments.
[9] See https://www.climatehubs.usda.gov/hubs/northwest/tools.

TABLE F-1 USDA Climate Change Hubs' Accomplishments During 2013–2023

Products	Metrics
Workshops and webinars	1,920 workshops and webinars with 142,474 participants
Presentations	2,077
Tribal engagements	608
Website visits	646,811
Publications	580 peer-reviewed publications and 1,184 white papers

SOURCE: See https://www.climatehubs.usda.gov/climate-hubs-quarterly-reporting.

3. **Outreach, Convening, and Training**: The hubs promote engagement, discovery, and exchange of information on climate science and climate-based tools to adapt to and mitigate the effects of climate change. Such efforts involve engaging with farmers, foresters, and land managers, often going directly to where they are located. The hubs develop products such as fact sheets, science briefs, opportunities for peer-to-peer engagements, workshops, webinars, and multimedia products. An example of this work is the Northwest Adaptation in Action profiles,[10] which highlight the climate resilience work of farmers, foresters, ranchers, and land managers implementing climate change adaptation and mitigation practices on the ground.

The climate hubs also work outward, providing climate change information, support, and resources to all levels of users, from schoolchildren to producers and land managers. Access to this information is provided through each hub's website. This information includes assessments, tools, webinars, demonstrations, and other didactic materials. Many of the communities with which the hubs work are further covered under the Justice40 Initiative, building resilience to extreme events and providing materials to help increase food security, support mental health, and more (USDA, 2024).

The climate hubs complement the federal network of climate science and information centers of the Department of the Interior Climate Adaptation Science Centers, NOAA Climate Adaptation Partnerships program (formerly known as Regional Integrated Sciences and Assessments), and others (see descriptions above) in directing audiences to usable, trustworthy, regional data and climate forecast services for incorporation into individual and community hazard and climate adaptation planning.

Examples of climate hub engagements with constituents over the last 10 years at the national level are presented in Table F-1.

---

[10] See https://www.climatehubs.usda.gov/hubs/northwest/topic/northwest-adaptation-action.